Breast, bottle or both

How to feed your baby, **your** way

Cathy Sage IBCLC

Cathy Sage

Copyright © 2024 by Cathy Sage

All Rights Reserved. No part of this publication may be reproduced, distributed, or transmitted in any form or by any means, including photocopying, recording, or other electronic or mechanical methods, or by any information storage and retrieval system without the prior written permission of the publisher, except in the case of very brief quotations embodied in critical reviews and certain other non-commercial uses permitted by copyright law.

Dedication

New parents,
you deserve our kind support

Acknowledgments

With heartfelt thanks to the wonderful parents who generously shared their stories and photographs in the creation of this book.

An additional huge thank you to my family, friends and colleagues who gently supported and encouraged me. Special mentions to Tammy Marlar for her wonderful photography, Nikki Wellspring for designing the beautiful cover, Lloyd Collins for stunning formatting, Fiona Sawyer for keeping me focused and Paul Cockerton for tireless editing.

Contents

Introduction 7
Feeding decisions 12
Getting prepared 42
Your baby's first breastfeeds 52
Positioning and attachment 60
Feeding quantity, duration and frequency 80
Settling your baby 112
Food, drink and medications 130
Breast and nipple challenges 142
Milk supply issues 168
Concerns about your breastfed baby 192
Feeding twins or triplets 224
Special situations 242
Breastfeeding away from home 272
Expressing and storing breastmilk 282
Bottle feeding basics 314
Preparing your baby's bottles 328
Your baby's first bottle feeds 340
Common bottle feeding concerns 354
Bottle feeding issues 362
Baby feeding and your mental health 370
About the author 418
Sources of support 420

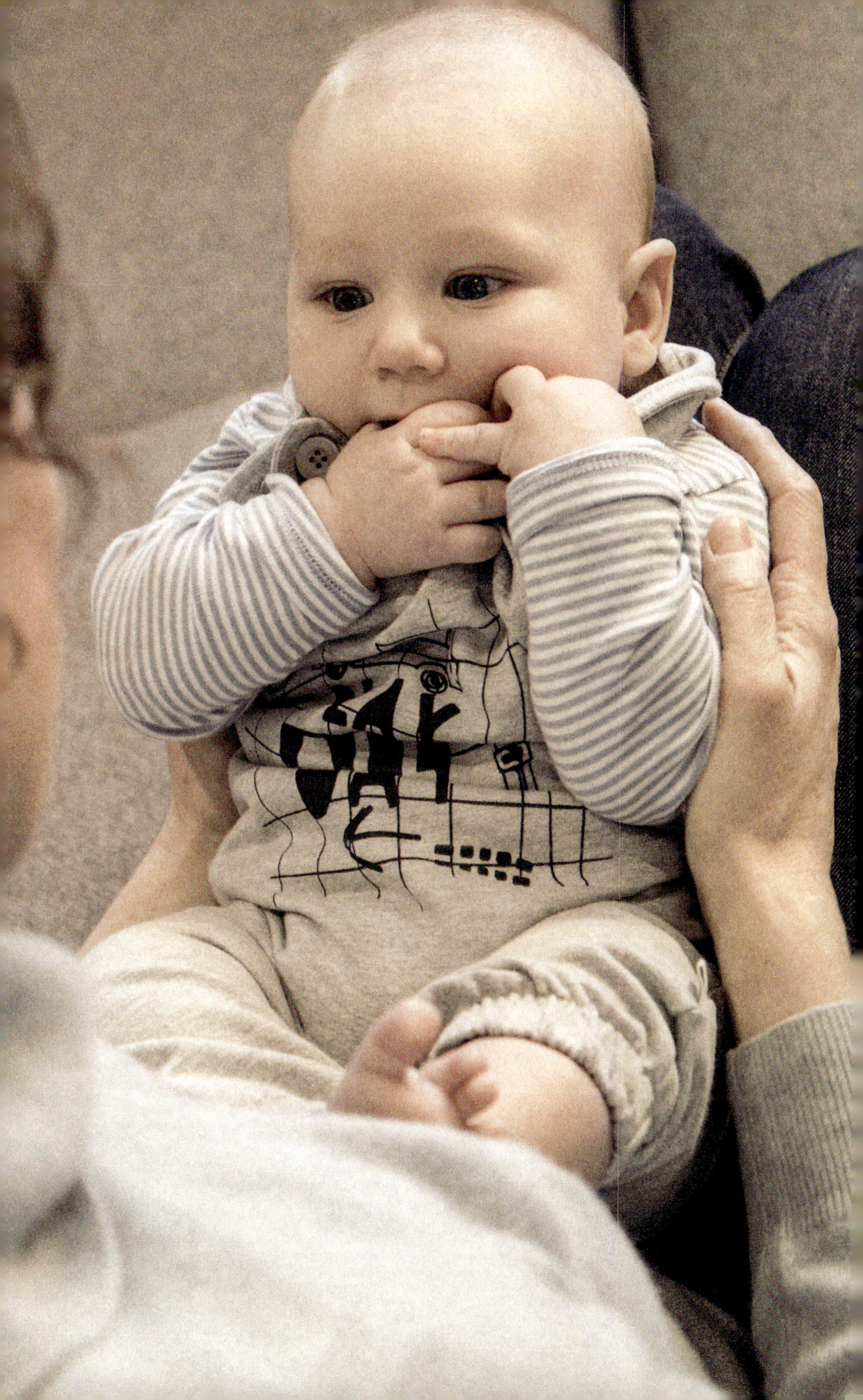

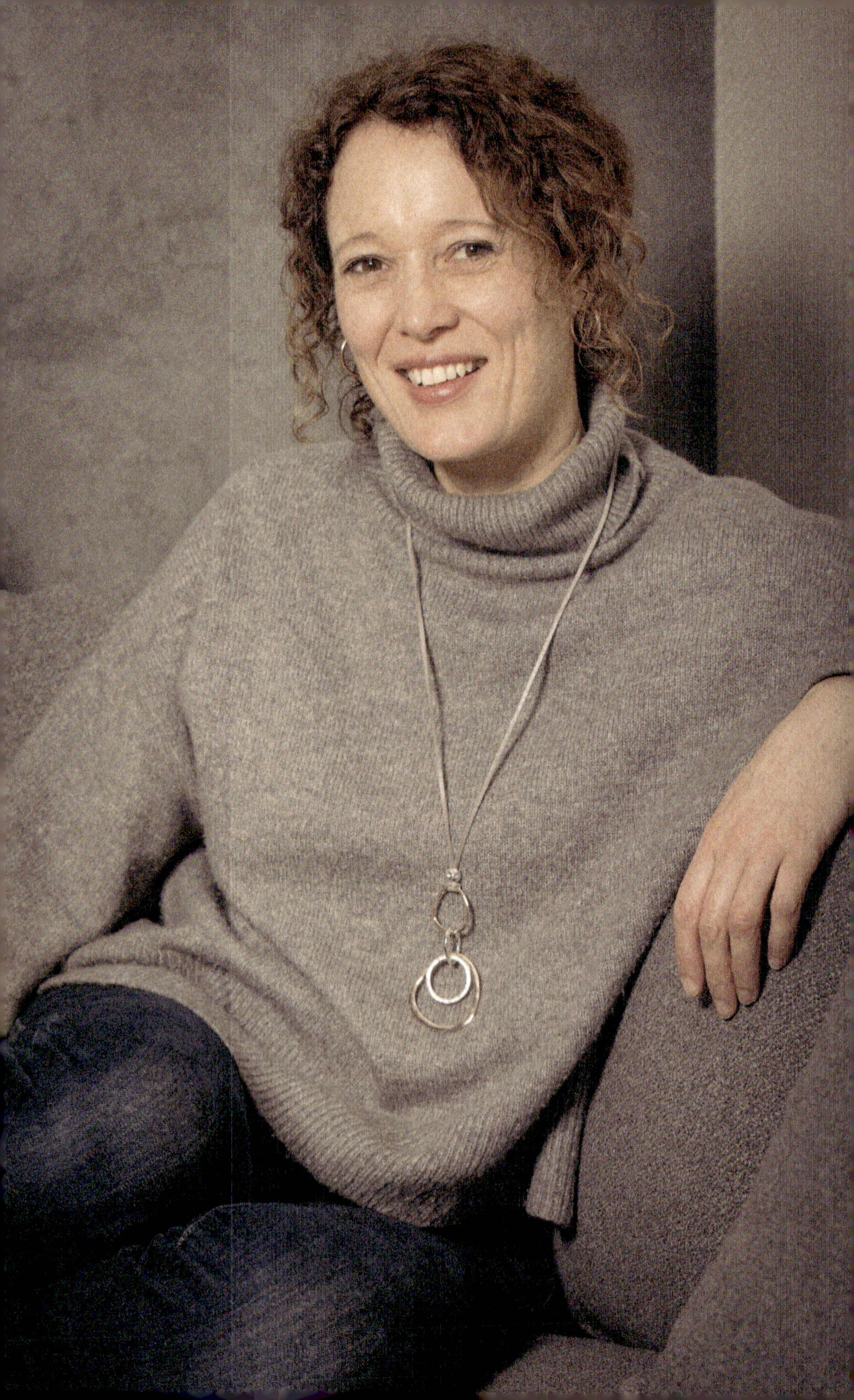

Introduction

Hi, I am Cathy Sage. I am a qualified and experienced baby feeding specialist. I am here to help you feed your little one comfortably and confidently. I have no agenda, apart from offering you the empathetic support that you need and deserve.

Over the last eighteen years I have helped many thousands of parents to overcome their breast and bottle feeding challenges. This book aims to answer the questions that I get asked so often in my work. It is the next best thing to chatting together on your sofa or by your hospital bed.

You and your baby are unique, and this guide can only be generic. Your midwives and health visitors, GP and hospital staff remain responsible for your care. Please reach out to get tailored support for your specific situation.

My qualifications

I am an International Board Certified Lactation Consultant (IBCLC), Certified Breastfeeding Specialist and fully qualified NCT Breastfeeding Counsellor and Antenatal Educator.

While there are no formal qualifications in bottle feeding, I have trained extensively with recognised experts in the field. I manage a busy private practice supporting new parents in baby feeding via home visit or online. I volunteered for years in two busy NHS hospitals and a GP surgery, and still run local services for the NCT. I have assisted thousands of new parents from their pregnancy and birth to two or more years afterwards.

My background is in science. I have a BA from Oxford University and MSc from Imperial College, London.

Choice of language
I have addressed this guide to the mother of the baby, answering her questions as I would if we were chatting.

My intention is to be inclusive, please forgive me if the word mother does not feel right to you, or if you are reading this as another wonderfully supportive person in your baby's world. In my practice I am delighted to work with everyone including parents who are adopting, those whose baby is being born with the help of a surrogate, gay parents, trans parents, those in any type of relationship and those parenting on their own.

In this guide I refer to mother/mum, partner, and baby (using the pronouns they/them). I assume one baby except for in sections exclusively for multiples. Please forgive any clunkiness in the grammar.

Why I wrote this book
I want to support you.

I am sure that you want the best for your baby. Perhaps you are overwhelmed with information and opinion from healthcare professionals, people selling books and services, manufacturers of baby stuff, family, friends and social media.

Most people and organizations have an agenda, even if it is not explicit. The NHS and breastfeeding organisations promote exclusive breastfeeding. Formula, bottle and pump kit companies are not benign, they need to sell you products. Some sleep consultants offer you routines and promise more control. Family members mostly want you to be rested and happy and often they would love to get involved in feeding your baby. Anyone you know is likely to try to sway you into their way of doing things as "the right way" even when it is not necessarily evidence based or right for you.

There is a distinct lack of actual scientific evidence on nearly everything to do with feeding and parenting. As a scientist by background, this drives me crazy! We must use logic alongside our parenting instincts and do the best we can. This book is based on the best evidence I can find at the time of writing, plus many years of practical experience. Feeding your baby can and should be a beautiful, bonding experience. However you feed your baby, this is the guide for you.

Many new parents feel a sense of loss when they do not have the birth and feeding experience that they hoped for. This is natural, normal and also hard. Please seek out some supportive family, friends and even therapists who can work through this with you and leave you free to enjoy your little one.

As a side note: I have taught antenatal classes for 18 years. If you are part of an antenatal class, I promise you that in every group almost everyone intends to mixed feed their baby, and some already know that they will formula feed. Very rarely, someone will have decided to give only breastmilk, and they might feel judged about that. People often don't talk about their feelings and decisions until someone is brave enough to open the subject. Most people will feel enormously grateful if you are honest about your experiences and totally accepting of theirs.

How to use this guide

The Guide is organised into chapters that cover your commonly asked questions. I answer with a high-level summary and then more detailed discussion.

In the section on **Feeding decisions** we discuss what type of feeding is possible and desirable for you. We consider breast and nipple size and shape, and think about medical conditions that can make breastfeeding more challenging.

Getting prepared introduces some of the thinking and products that you might need in advance for breastfeeding, expressing and bottle feeding.

Your baby's first breastfeeds discusses feeding in the early hours with your new baby. We then dive into details of **Positioning and attachment**, learning how to latch your baby comfortably in ways that work for you.

In the section on **Feeding quantity, duration and frequency** we look at signs your baby is feeding well and consider everything from when to swap sides to the feasibility of routines. Most of this information applies to breast and bottle feeding babies. We take a closer look at **Settling your baby**, and then tackle questions on mothers' **Food, drink and medications** when breastfeeding.

As women sometimes find breastfeeding difficult, we have a section on troubleshooting **Breast and nipple challenges** covering many areas from soreness to mastitis and nipple thrush.

Many mothers are worried about their milk supply. In **Milk supply issues** we look at reasons for high or low supply and how to manage them.

The section **Concerns about your breastfed baby** considers your common questions from spitting up to tongue-tie.

Parents of multiple babies have a dedicated section on **Feeding twins or triplets**.

In **Special situations** we consider the feeding implications of medical challenges such as the baby being in NICU (Neonatal Intensive Care Unit) or the mother having severe blood loss.

We have a section all about **Breastfeeding away from home**, including thoughts about what to do when returning to paid work.

Most mothers pump milk at some point. Common questions and concerns are covered in **Expressing and storing breastmilk**.

We delve deeply into **Bottle feeding**, starting with the basics of how to choose a bottle and formula, and how to safely **Prepare your baby's bottles**. We cover **Your baby's first bottle feeds**, and **Common bottle feeding concerns**. There is a section dedicated to overcoming **Bottle feeding issues**.

The final section addresses **Baby feeding and your mental health**, talking honestly about the adjustments and realities of becoming new parents.

We end with **Sources of support**.

Feeding decisions

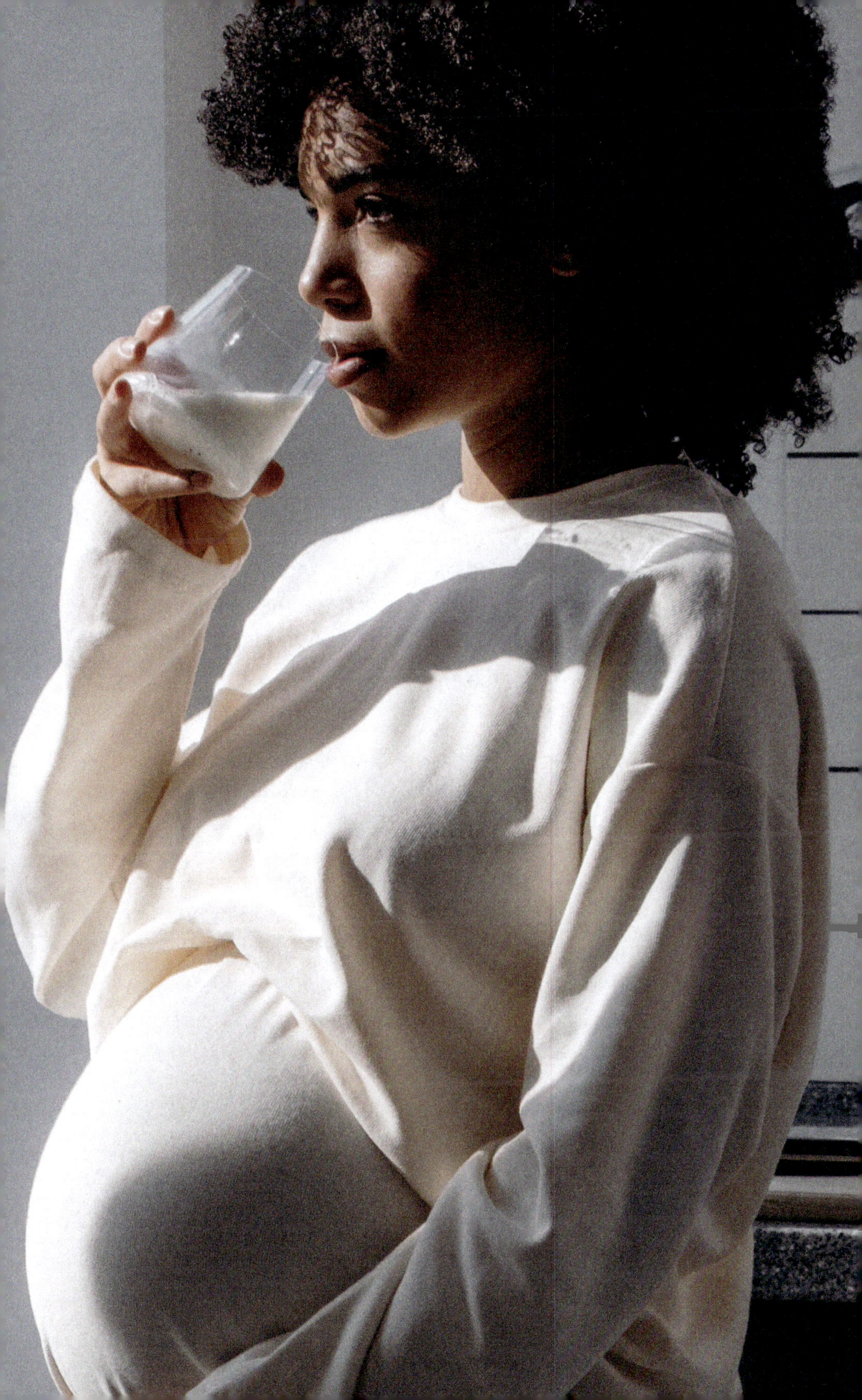

Is breast really best?
Clinically, yes, but formula is also fine.

If you are reading this while expecting your baby, you might be thinking about how you intend to feed your little one. Perhaps you have a clear view about breast or bottle, or maybe a combination of the two. Alternatively, you might just want to wait and see how it goes and how you feel at the time.

A little bit of science.

Human milk is a fascinating, dynamic substance that changes with your baby's needs. It actively supports your growing baby's brain, body and immune system, and contains sugars to feed their colonising gut bacteria. Antibodies and viral fragments prime your baby's immune system, and you act as a real time virus checker, passing on protection to your baby if either of you becomes unwell. Breastmilk composition changes through time as your baby goes through toddlerhood and beyond.

Breastmilk is free, available on demand, and delivered immediately at the perfect temperature. It is an environmentally sustainable option.

We know that mothers who breastfeed are slightly less likely to get some forms of breast cancer, and that breastfed babies are less likely to be hospitalised with stomach upsets or respiratory infections. Breastfeeding your baby can feel wonderful, as you are both filled with hormones that help you to be relaxed and happy.

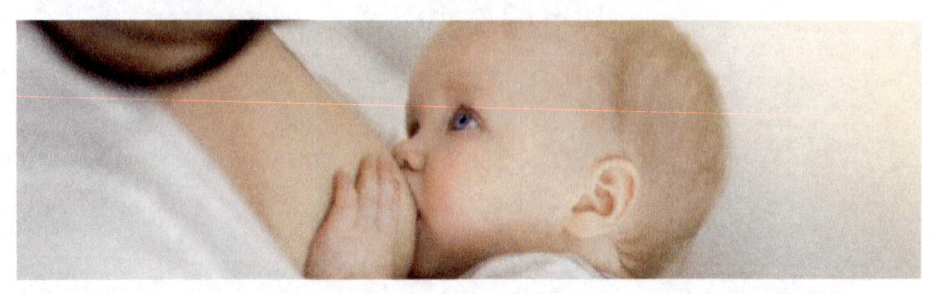

What is in breastmilk?
- Water.
- 1000+ types of proteins.
- Long chain fatty acids.
- Carbohydrates, specifically.
- 200 types of sugars for energy and to promote suitable gut bacteria.
- Vitamins and minerals .
- Millions of live cells .
- Antibodies and viral fragments.
- 40+ enzymes.
- Growth factors.
- Hormones.
- 1,400 microRNAs.
- Anti-inflammatories.
- Lots more!

What is in formula milk?
- Water.
- Proteins (derived from cow's or goat's milk).
- Fat (mainly vegetable oils).
- Carbohydrates.
- Vitamins and minerals.

Additional components added by the manufacturer that might make the formula more like breastmilk but are not required by regulation.

Formula milk is nutritionally complete. Its composition is strictly regulated under UK Law. A huge amount of research is undertaken every year to try to make formula as close to breastmilk as possible. We know that babies thrive on formula milk, and that they continue to grow and develop normally. There is much more detail in the section **How to choose infant formula**.

However, it would be wrong to suggest that formula is the same as, or just as good as human milk.

We know that published research is likely to be biased towards the benefits of breastfeeding. Studies that show the benefits of formula feeding are often not put into journals. There have been sibling studies of thousands of families with one breastfed child and one formula fed child, which find very few significant differences between them in long term health outcomes.

We live in a culture where a mother is expected to somehow minimise all risks and do her absolute best for her baby. There is no question that breastmilk is the ideal milk for a human baby, so pressure to breastfeed is laid on the mother. When breastfeeding is not possible or desirable the mother can suffer from anxiety, depression and feelings of guilt. However, we all need to recognise that many women cannot make milk, or sufficient milk to exclusively breastfeed. It might not be a woman's choice to breastfeed for a host of complex reasons that are not visible to those forming judgement.

From a practical and logistical point of view, breastfeeding is a full-time job. Women need adequate physical, emotional and financial support to commit to it. Pressures of paid or unpaid work may make it almost impossible to breastfeed exclusively or even partially. We explore some of these themes in more detail later in this guide.

It is important to remember that a mother needs to invest in herself as well as in her baby, so that everyone is OK. Many families are going to work best overall when they use formula milk for their baby, either entirely or in combination with breastmilk.

The key message is to feed your baby, enjoy your baby, and try not to let the first get in the way of the second.

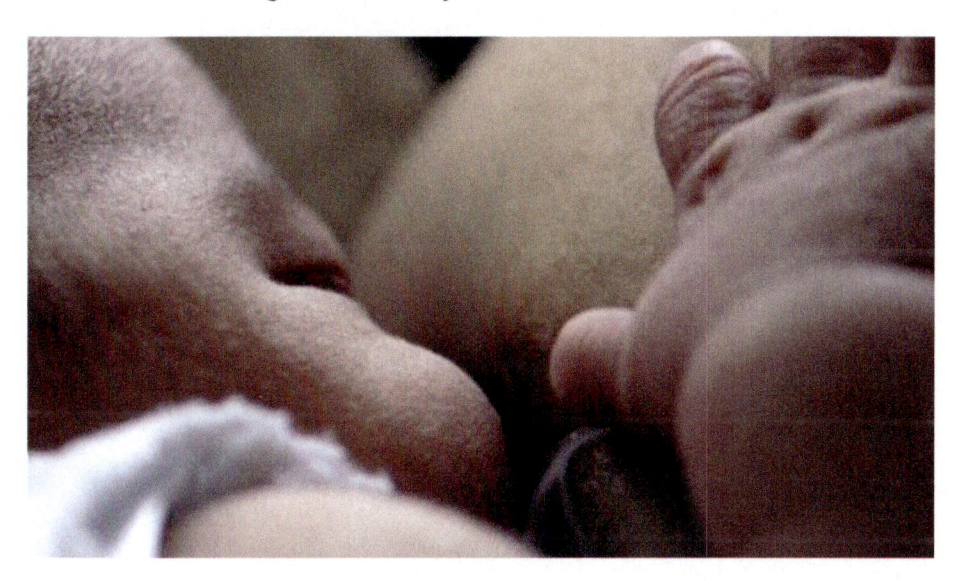

When do I have to decide about breast or bottle feeding?

Some women will have made their feeding decisions during pregnancy (or before), and others wait and see how they feel about it at the time.

Many mothers use a combination of breast and formula milk right from the start, and we talk more about that later in this guide. If you are undecided, why not buy the minimum kit that you will need for bottle feeding (a couple of bottles and some ready to feed number 1 formula) and see how you feel? It does not have to be a binary decision, and there is no rush to make it.

Take one day at a time.

If you are certain that breastfeeding is important for you, think about what support you might need in the early days. You may wish to research local feeding drop ins, lactation professionals and support lines. Let your friends and family know how much you would appreciate their help in honouring your decision, and looking after you as much as possible while you get started on your breastfeeding journey.

If you are planning to bottle feed, research which bottles, sterilisers and formula feel right for you and your family. Think about who you might like to get involved with feeding your baby, and perhaps watch and read some guidance together on paced and responsive bottle feeding techniques. There is more about all of this later in this guide.

Are there health or breast concerns that might impact on my feeding decisions?

Most women can breastfeed, but some medical conditions and medications for a woman or her baby may prevent exclusive or even partial breastfeeding.

Mothers might be worried that they will not produce breastmilk, or not enough milk for their babies. The good news is that most women can breastfeed their baby with appropriate support.

However, we know that some women cannot breastfeed, or fully breastfeed, and it helps to get informed about this before you start.

For breast milk production the breasts need to have sufficient glandular tissue and nerves that work well, and the correct hormones need to be produced in sufficient quantities. To maintain production, breast milk needs to be removed by the baby or a pump regularly and effectively.

This means that women are more at risk of low supply if they did not grow breasts during puberty, or their breasts are asymmetric or unusually shaped. Some illnesses and medical conditions such as breast or brain cancer, and treatment for some childhood cancers, can have an impact on the breasts or hormones that control them.

Breast injury or surgery can clearly influence milk production, particularly breast reduction where glandular tissue is removed and nerves and ducts damaged.

Hormonal issues might be indicated by polycystic ovaries, known thyroid conditions, or perhaps undiagnosed infertility requiring IVF. There are few other health conditions and medications which might prevent a woman from breastfeeding, these are explored later in this guide.

Some medical conditions might make it harder or impossible for your baby to breastfeed, and you will be advised about these from the medical professionals working with you.

If you are concerned about any of these points, please consider contacting an experienced lactation consultant for support. In many cases when a woman wants to breastfeed this is still possible to a greater or lesser degree. We work with what we have got, celebrating any breastmilk that is available, building milk supply as much as possible, and supplementing with as much infant formula as baby needs.

Does the size of my breasts matter?

Most women produce sufficient milk for their baby regardless of the size of their breasts.

If your breasts developed normally in puberty, and you noticed changes in pregnancy, chances are that there is plenty of glandular tissue to make milk for your little one. Bigger breasts do not always have more glandular tissue.

The amount of milk that breasts can store between feeds varies from woman to woman, and you will get to know your own body's milk supply and storage capacity when the time comes. Women with a smaller capacity to store milk may need to feed more frequently than others.

Differences in physiology mean that women often experiment with the most comfortable feeding positions for themselves and their babies. We look at this more fully in the following chapters.

For breastfeeding support after breast surgery, a good resource can be found at the Breastfeeding After Reduction Surgery website.

What about my nipples?

In most cases, women who wish to breastfeed can do so.

Nipples come in many shapes and sizes. They may protrude, be flat or sometimes inverted. They may be smooth or have the occasional hair around them. They can be pink or brown or pale. Some are unusual, I have seen a few like button mushrooms. Women may have chosen to tattoo or pierce them.

If you are particularly worried about your unique situation, it can help to chat to your midwife or a lactation consultant in advance of the birth, to help you to feel as confident and prepared as you can be.

Can I breastfeed with pierced nipples?

Yes! Please remove your nipple jewellery before breastfeeding.

Jewellery can easily get dislodged or damage your baby's mouth. Most people take it out for the whole time they are breastfeeding their baby, others do it feed by feed. Obviously, keep your hands and the jewellery clean when taking it out and reinserting it.

If you must keep it in place (not recommended) make sure that it is not loose or broken before each feed and observe your little one carefully.

You might leak a bit of milk out of the piercing holes, laid-back positions can help if the flow is a bit overwhelming for your baby.

What if you want a new piercing? As ever, choose a reputable piercer! Most will not give you a new piercing if you are pregnant or breastfeeding as the holes can take months to heal. Wait three or four months after you have stopped breastfeeding before getting a new nipple piercing.

Is it OK to breastfeed if I have a tattoo?
Yes.

You might be thinking about whether it is safe to breastfeed if you have a tattoo, or to get a tattoo while breastfeeding. To be honest, there is not sufficient research out there to be totally sure. It would be reasonable to suggest that the health benefits of breastfeeding are likely to far outweigh a theoretical risk from having a tattoo.

Most reputable tattoo artists will not knowingly tattoo a pregnant or breastfeeding woman. This is for liability reasons: it is possible that the mother will get an infection from the tattoo process or during the healing period. We do not know much about how the tattoo inks break down in the mother's body in the years after a tattoo is placed, or the extent to which these inks find their way into breastmilk. As far as we know the chance of ink being in the breastmilk, or impacting on your baby, is incredibly small.

When can I get a new tattoo?

Common sense advice is to wait until at least six-nine months after birth, when your baby is no longer totally dependent on breastmilk.

As always, minimise the risks of infection by carefully researching your provider and opting for someone who follows all the safety guidance. Keep your new tattoo clean with mild soap and water, do not pick any scabs, and keep it out of direct sunlight while it heals.

Is it safe to have a tattoo removed while breastfeeding?

When a tattoo is removed, lasers break the inks into smaller particles that go into the immune system and get cleaned out by your liver. The main immediate issues are pain, inflammation and possible infection. We do not know about whether the ink particles enter breastmilk or what impact that might have, but the theoretical risks are very small. It would make sense to breastfeed your baby for as long as you both want, and get the tattoo removed after your baby is weaned from breastmilk.

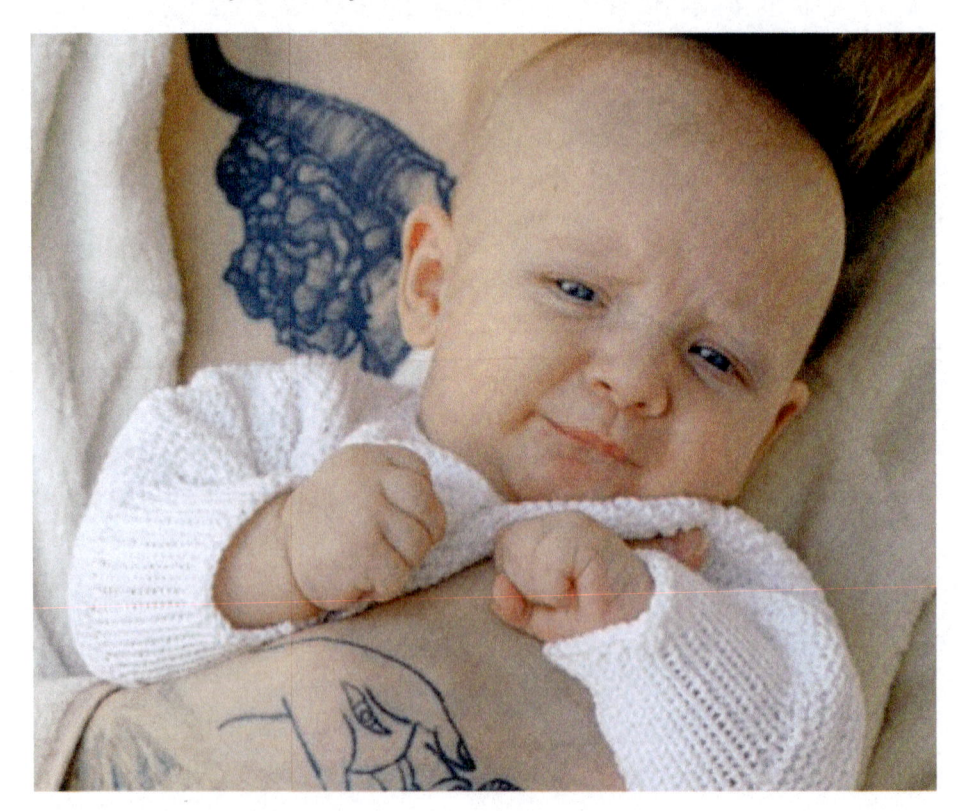

My breasts are different, can I breastfeed?

Most breasts produce at least some milk. If you are worried, please ask a professional.

Women often secretly worry that their breasts are not normal, or right for breastfeeding. In my practice, I have seen many thousands of breasts, and very few of them are abnormal! They can have a range of cup sizes, nipple placements, nipple shapes and sizes. They can be naturally a little different each side. If your breasts changed appropriately during puberty, and again while pregnant, it is likely that they respond well to your hormones and will be fine for breastfeeding.

I am more concerned when I meet a woman whose breasts or medical history show some issues that might indicate that there is insufficient breast tissue to fully breastfeed.

These women might have:

- Not grown breasts as a teenager.
- Grown asymmetric breasts when one is significantly bigger than the other.
- Breasts that look "tubular" or narrow.
- Areolas that look puffy or swollen.
- Widely spaced breasts with more than a hand's width between them.
- Experienced their milk "not coming in" or a very low supply.
- Had corrective breast enhancements or reduction surgery.
- Undergone radiotherapy.

These diagrams show mammary hypoplasia, or insufficient glandular tissue. With acknowledgment to Circlo Imagen para el Diagnóstico.

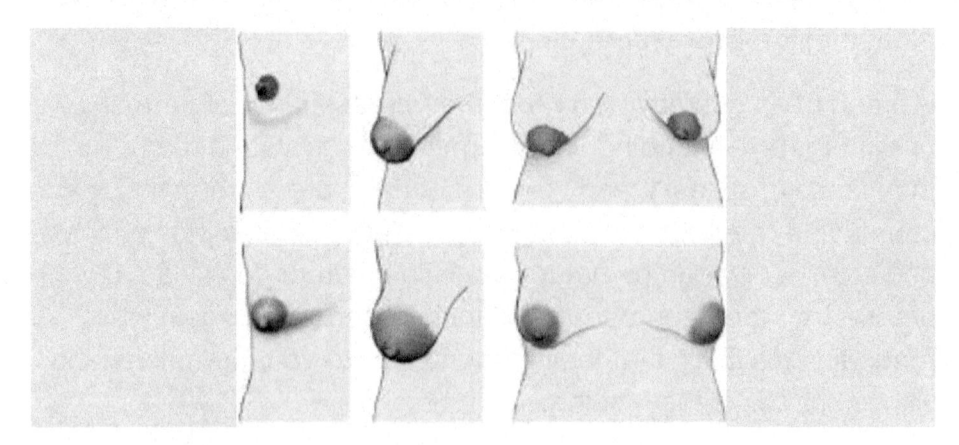

If you suspect or know that your breasts might be an issue for feeding, please seek out skilled support. You might enjoy some breastfeeding while also topping your baby up with some formula or donor milk. Please see the chapter on troubleshooting milk supply issues.

Some women with insufficient glandular tissue will almost fully, or completely, formula feed their little one. These medical conditions are nobody's fault. Reach out for skilled support if you would like to talk about how you are feeling. I have met three or four women through my years of practice whose medical practitioners and midwives had not noticed, or thought to mention, their mammary hypoplasia. When I tried to find images online, there were none. When you think about just how many breasts we can see online, I feel that this is letting women down.

I have functioning breasts, but I am not pregnant. Can I make milk?
Yes.

Some women choose to induce lactation, for example when adopting a child, working with a surrogate or when intending to co-feed a child alongside their partner. There are different ways to induce lactation. It takes huge dedication to build up a milk supply, with many hours of pumping with a machine several times a day for several weeks.

In some protocols, the woman inducing lactation takes birth control pills for a while to simulate pregnancy before stopping them and starting the pumping regime. I would recommend seeing an experienced lactation professional while embarking on this incredibly loving and challenging journey.

Can I share breastfeeding with my lactating partner/friend/family member?

I have worked with several women who co-feed their children, and it can be a beautiful way for the babies to bond with their caregivers.

In some cases, co-feeding is necessary when one parent cannot produce a full milk supply. In some of the lesbian couples I work with, it is just gorgeous that either parent can feed the little one (and possibly older siblings) whenever they want.

Some couples choose to have babies at the same time. Alternatively, one partner has a baby first and breastfeeds, and the second then also has a baby.

When co-feeding it is important to remember supply is linked to demand, breasts produce milk according to how much is taken out by the baby or a pump. Sharing feeding between two or more people may result in each lactating person producing less milk than they might otherwise (unless they also express milk to boost supply). Of course, both the nursing adult and baby get much more from breastfeeding that milk; it is a wonderful, healing and loving experience regardless of how much milk is being produced. As normal, keep an eye on how your baby is doing.

I'm pregnant and trans, can I chest or breastfeed?

Breastfeeding is possible, the milk volume will depend on your hormones and the amount of functioning breast tissue. Please find a supportive IBCLC to work with you.

My trans friends and clients often face additional challenges when it comes to parenting and feeding their children. For a trans masculine person, choosing to become pregnant often requires coming off hormones. It is a profound decision, and many people struggle with the significant body changes and associated dysphoria. It can be difficult to deal with the physical and emotional impacts of former medications and possible surgeries. On top of this we have social expectations and prejudices.

If you have milk producing tissue and you want to chest or breastfeed, you should be supported to do so.

If you've had top surgery the amount of milk available will be a function of the glandular tissue remaining, the type of surgery you had (and how it impacted nerves, ducts and blood supply), and how long since it was performed. Supplementation with donor or formula milk will almost certainly be necessary. You can do this with a nursing system on your chest (see later sections) or with a bottle. Every millilitre of colostrum or milk you produce will be beneficial to your little one.

If you bind your chest, you might be at a higher risk of developing blocked milk ducts or mastitis. You might consider a larger binder for the time that you are feeding your little one.

Testosterone interferes with prolactin, one of the hormones needed for making milk. Taking testosterone may cause a significant decrease in your milk supply. However, you can still use a nursing system and have a feeding relationship with your little one.

We also know that holding your baby on your chest skin to skin or lightly dressed can be a wonderful way to keep your baby warm, comfortable and secure regardless of how the feeding happens. Equally, if this does not feel right for you, that is OK too.

It would be great to work with a supportive IBCLC who can help you to figure out how much milk is available and how much to top up if required. You will also benefit from a network of like-minded friends. The book "Where's the Mother" by Trevor MacDonald is a moving account of new parents going through this experience.

I'm a trans woman and take hormones. Can I make milk?

Limited milk production is possible and could be used alongside significant supplementation with donor milk or formula. This is an under-researched area and is often overlayed with unkind societal judgement. You need and deserve some specialist support.

If you have been on hormones for some time, you might have developed some glandular tissue that could make milk. It is possible to encourage this process using a similar protocol to any non-pregnant person inducing lactation in advance of meeting their baby. Scientific evidence around this is very limited. Please seek out a supportive healthcare team to look after your physical and emotional wellbeing, and to help monitor your baby to determine supplementation amounts and frequency.

Babies receive more than milk when they are held close to their parent, and all close and loving contact is beneficial. You might like to consider a supplemental nursing system that would allow you to chest feed, perhaps in tandem with a nipple shield. You will find more in the section "I do not want my baby to have a bottle, what are my other options?"

Can I breastfeed after breast surgery?
Most likely yes! But you might also need to top up.

Most women who have had breast surgery still produce some breastmilk. The amount depends on many things including the reason for the operation, type of surgery, the way it was performed, and when it was done. This section covers implants, reductions, biopsies, lumpectomies and mastectomies.

If a mother wants to breastfeed, we should celebrate any breastmilk that is possible alongside supplementing with formula as necessary.

Breastfeeding with implants (breast augmentation)

Breast implants are used to increase the size of the breasts or for reconstruction after a breast or part of a breast has been removed.

If the breasts contained normal breast tissue before having implants, they may not impact a woman's ability to breastfeed. Implants tend to be inserted from incisions at the sides of the breasts, away from the areola, and behind the breast tissue. Implants positioned directly above the chest muscle are more likely to put pressure on the glandular tissue and block milk flow than implants positioned under the chest muscle. Frequent feeding and/or expressing can help to reduce the feeling of fullness that might impact on milk supply.

Some women get implants as reconstruction after surgery or because of underdeveloped or misshapen breasts on one or both sides. In these cases, there may not be sufficient breast tissue available to make much milk for their baby.

Over the years I have met several women whose babies were not thriving or putting on weight, and nobody had asked about the development of their breasts in puberty or about corrective surgeries. I find that many health professionals simply do not mention it in conversation, or do not fully read a woman's notes. This can leave a woman feeling like she is failing somehow, which could not be further from the truth.

Babies need to be fed and loved, and it does not really matter how they are fed. If a woman has medical or physiological issues that affect her breastmilk production, or has chosen a course of action in her past that perhaps she now regrets, she needs support and understanding. She should not judge herself, her body, or feel judgment from anyone else.

Breastfeeding after breast reduction

Breast reduction decreases the size of the breasts, and often results in lower breast milk production. The process involves removing breast tissue and excess skin, and very often repositioning the areola by cutting around it. Moving the nipple during breast reduction surgery can injure the milk ducts and affect the nerves and blood supply to the area. Over time, nerves slowly grow back and breast tissue can heal to an extent. If the surgery was performed without cutting the area around the nipple and areola, there is a better chance that breastfeeding will be successful.

In my experience, many women will produce some breastmilk after a breast reduction, but most do not make sufficient milk for the baby even with frequent nursing and expressing. It is likely, and desirable, to supplement the breastmilk with formula top ups by bottle.

Breastfeeding After Reduction (BFAR) has an active website and social media presence for anyone needing more information and support: www.bfar.org

Breastfeeding after mastectomy

A mastectomy is the removal of a breast. They may be done because of breast cancer, or some women choose a mastectomy due to very high genetic risk of developing cancer.

After a mastectomy, the breast involved may not be able to produce much milk, if any at all. It depends on how much of the breast tissue is removed and whether any additional treatment is needed. Radiation therapy can also cause further damage to any remaining breast tissue.

When one breast is removed but the other is not affected, it is sometimes possible to make enough milk for a baby from just one side without the need for supplementation.

Breastfeeding After a Lumpectomy

Women may have lumpectomy surgery to remove a breast lump. A lumpectomy may damage milk ducts, tissue and nerves in that area, and it is likely that this will affect breast milk production in the damaged part of the breast. Less invasive lumpectomies not involving an incision near the areola will probably only have a minor effect. Sometimes radiation therapy is also required, and this can further damage breast tissue.

Breastfeeding After a Breast Biopsy

Sometimes it is necessary to surgically remove a piece of breast tissue to check for cancer or infection. This is often minor breast surgery and generally does not affect the ability to breastfeed. Feeding might be more challenging if the incision was extensive and near the areola and nipple. Incisions that are orientated towards the nipple, like the spokes on a wheel, are less likely to cause nerve and duct damage.

A needle biopsy is performed by inserting a needle into the breast to remove the contents of a breast lump, cyst or abscess. This procedure rarely has a negative effect on milk production or the ability to breastfeed.

Breast lift surgery

Mastopexy repositions the breast tissue to reduce sagging, and implants are sometimes added at the same time to increase fullness. Breast lift surgery by itself may not always impact milk production because no or limited glandular tissue is removed and the incisions are usually not deep enough to affect critical nerves. It makes sense to work with an experienced lactation professional in these instances.

Tips for breastfeeding after breast surgery

- Contact the surgeon who performed your breast surgery to find out how the procedure was carried out and if it might interfere with your ability to breastfeed.

- Talk to each midwife, obstetrician, your GP and the baby's paediatrician about your desire to breastfeed and let them know that you have had breast surgery. Please do not assume that they have read your notes.

- Consider attempting to hand express colostrum before your little one is born.

- Tell the hospital midwives and feeding supporters about your breast surgery and make sure that they carefully monitor your baby during your hospital stay.

- Start breastfeeding your baby as soon as possible after delivery. Keep baby skin to skin if possible and try to feed very frequently, at least every 2-3 hours, to help build up your milk supply. If you can, hand express or use a breast pump for a short time after each feed to ask your breasts to make more milk.

- Read the section on "How can I tell whether my baby is getting enough breast milk?"

- Focus on your baby's behaviour and what is coming out in their nappy.

- Seek specialist feeding support from a local IBCLC qualified lactation consultant.

- Take your baby for regular weight checks as indicated by your health visitor.

Is bottle feeding easier and more predictable?
Sometimes.

Bottle feeding allows other people to get involved. You will have a clearer idea about how much milk your baby is drinking. Many babies will take bigger feeds by bottle than they might at the breast, and this could make a feeding routine possible. Parents of multiples often rely on some bottle feeding.

For some people, bottle feeding is easier than breastfeeding. However, it takes considerable time and effort to wash, sterilise and prepare baby bottles, and more forethought is required, for example when travelling.

Bottle feeding can allow a woman time for rest and recovery, which is especially critical if she has a preexisting health condition such as some forms of epilepsy or multiple sclerosis.

I work with many women who are single parents by choice or necessity. They may choose to bottle feed so that other people can support them, especially if working with a maternity nurse or night nanny.

Can I feed my baby with breast and bottle?
Mixed or combination feeding is offering both breast and bottle. The bottle could contain breastmilk or formula.

Most parents I work with plan to combine breast and bottle feeding, ideally from the start. Almost every mother wants their partner or another adult to be able to feed the baby, to give them a break and allow others to bond with the little one.

In many situations mixed feeding is necessary for the baby or babies to get the calories they need to thrive. It can be a temporary or long term solution and means that babies can receive the benefits of some breastmilk.

Common scenarios include:

- Baby cannot fully breastfeed because they are unwell, premature, or have problems with their mouth or tongue.
- Mother cannot fully breastfeed due to a health condition, breast or nipple problems or surgeries, or personal choice.
- A mother has multiple babies.
- Other people want or need to be involved with feeding the baby.
- Mother and baby need to be separated.
- The woman is building a milk supply through induced lactation, re-lactation or sharing feeding with another person.
- A birth parent is breast or chest feeding when it feels possible. Dysphoria, eating disorders, depression or neurodivergence can make it much more challenging to exclusively breastfeed, but some might be tolerable.
- A previously traumatised parent is breastfeeding for some of the time when they can, for example during the day.

Mixed feeding using formula means that the mother's breasts are not used as much or as frequently as they would if she were fully breastfeeding. Her milk supply will go down to match the demand. In the first few days and weeks after the baby's birth, this can have a considerable impact on overall milk supply. Some women's supply declines to the point where they decide to no longer breastfeed.

However, well managed mixed feeding can be great for the whole family. After about a month, a woman's milk supply is more established and flexible. A common situation is for a partner to offer a bottle feed of expressed milk or formula before they go to bed, allowing a mother to go to bed early and effectively miss a feed. Most women find they have a lot of milk in the early hours of the morning, so this tends to be a time when they actively need to feed or express milk.

There is not much clear advice around mixed feeding, and many parents feel left in the dark. As babies are getting both breast and bottle, it can be hard to know exactly how much milk they are having. Parents need to be responsive to what is going on for their baby, monitoring baby's behaviour, looking for feeding cues, and checking baby's nappy output. Many families happily mixed feed their baby for weeks, months or years.

When could I start mixed feeding?
Start whenever you need to.

There are many reasons why you might combination feed from the start. If there is no urgent need to mixed feed, hold off for about a month if you can.

If the breasts are functioning well, breastmilk is made to match the amount being taken out. If your baby is being offered formula, you will not make as much milk. This might mean that your milk supply will reduce to a level where your baby is no longer very interested in breastfeeding. Some women express breastmilk to keep their supply up.

Please reach out for some skilled support if you can. It is great to find a sweet spot where the balance of breastfeeding and formula is sustainable for you. If mixed feeding in the early days, it is important to know that not using the breasts for several hours in a day leaves them uncomfortably full and reduces milk supply.

On the other hand, expressing much more than your baby needs in 24 hours can cause an oversupply of milk. If there is no urgent need to start mixed feeding, it can be beneficial to the breasts to wait until breastfeeding is established after four to six weeks. This reduces the chance of over supply or undersupply of breastmilk, and the breasts seem to offer more flexibility.

If I give my newborn a bottle will it ruin my chance of breastfeeding?

There is no clear evidence that giving your baby milk from a bottle will make them reject the breast: most babies cope well with breast and bottle feeding.

Obviously, there is more to this question, and it very much depends what is going on for you and your little one.

If you want to breastfeed but it is difficult, please get some help as soon as possible. If your baby is struggling to breastfeed it would not be a surprise if they begin to prefer a bottle.

However, please do not hesitate to feed your baby by bottle while you get support with breastfeeding. You can protect your milk supply by getting milk out by hand or machine around your feed times (about every three hours).

Some people have concerns that babies might become accustomed to fast flowing bottles, and then have less patience at the breast. Paced or responsive bottle feeding with an appropriate teat can help to reduce this theoretical risk. Alternatives to bottles are available.

Can I purchase breastmilk or get it from a milk bank?

Many larger hospitals have milk banks for vulnerable babies, and it is sometimes possible to receive donated milk longer term.

There are many clear health benefits of offering human milk to babies, especially those born early or unwell. In some NHS hospitals, treated milk from screened donors is prescribed to babies in neonatal intensive care. In many larger hospitals it is also available to other babies that need it. If stocks are high enough, parents might be able to get donor milk for a full-term baby on the postnatal ward, particularly if the mother intends to breastfeed but finds herself temporarily unable to do so. This milk is donated and provided for free in an NHS setting. If you know that your little one is likely to be born early and/or vulnerable, talk to your midwives or hospital consultants about what might be possible. Most mothers make some milk for their babies, and staff in the neonatal unit will help them to feed it either directly or by other methods, there is more in this later in the guide. In some cases, human milk from a milk bank can be delivered longer term to babies being cared for at home.

In the past, sharing human milk was much more common. Family and communities were involved, and others chose the option of an unrelated "wet nurse" to breastfeed their baby. While less talked about these days, it is sometimes possible to enter into an informal agreement with a trusted friend or family member

37

who is lactating and willing to offer some milk either by direct breastfeeding or expressing. I have, for example, worked with families of two women where the non-birth mother fed her own baby and her partner's baby. I have also supported families of two dads, where one of their sisters provided expressed breastmilk.

In many instances lactating friends will step in if you ask them.

The key thing to remember is that human milk can carry some bacteria and viruses such as HIV, so it is essential that it comes from a trusted source. It is vitally important that expressed milk is collected and stored appropriately. Donor milk from a milk bank is screened and heat treated to reduce risks involved, but this does not happen in less formal arrangements. It is also a reason to think carefully before accepting milk offered from people you don't know but met online.

Some cultures are cautious about milk sharing as they class babies fed the same milk as "milk siblings." There are issues around, for example, marrying a milk sibling in later life.

If you are in an informal milk sharing agreement, especially if another person is directly breastfeeding your baby, do think about the emotions that might go along with it. There are bound to be mixed feelings when your baby is intimately connected to, or relying on, someone else. While it could be lovely to see your baby being adored by other people too, it would be possible to feel slightly excluded and jealous. You might also find yourself subject to other people's often ill-informed opinions. In any case, it would be good to keep an open dialogue about it with the milk donor (if known to you) and significant people in your and your baby's life. You might also appreciate some therapeutic support for yourself, which I would recommend to any new parents and especially to those in more complicated situations.

I have an older child that is still breastfeeding, can I continue to breastfeed when I am pregnant or have a newborn?

Yes, many mothers who are pregnant will continue to offer their baby or toddler breastmilk.

Bear in mind that some nurslings do not like the taste of the milk when their mother is pregnant, and some self-wean. You might also find that your breasts or nipples are tender, and you do not feel like breastfeeding.

Towards the middle of your pregnancy the milk starts to revert back to colostrum for the new baby, and again, some nurslings wean or mothers discontinue breastfeeding. You do not have to.

When the new baby is born, please ensure that they have access to the breast before the older nursling so that they can take all the milk they need and keep an eye on their nappy output and weight as you would with any baby.

Can I just give my baby colostrum and then not breastfeed after that?

Yes, of course you can! You decide what is best for you and your baby.

Colostrum is particularly high in antibodies that help to protect your baby and build up their immune system. It helps their gut to grow friendly bacteria and reduces inflammation. Colostrum is particularly great for premature, vulnerable and low weight babies. It also offers everything that a newborn needs in their early days, before they are ready to take bigger volumes of milk.

If you wish to give your baby colostrum and then not breastfeed, it is unlikely that you will have made problems for yourself in stopping. There is more about winding down or stopping breastfeeding later in this guide.

How long should I plan to breastfeed for, in terms of weeks or months?

This is entirely up to you and your baby.

The World Health Organisation encourages mothers to exclusively breastfeed until their baby is six months old, and then offer additional food alongside breastmilk until their baby is two years and beyond. There are definite health advantages of this approach for mother and baby, especially in countries without access to clean water to prepare bottles of formula milk. Human milk composition changes as your baby gets older and is especially valuable in giving them lots of antibodies to protect them as they explore the world.

However, most babies in the UK are exclusively breastfeed for a much shorter time than this. The majority are offered formula for at least some feeds when under six weeks old. There are numerous reasons for this. Generally, women feel that they stop breastfeeding before they wanted to. There is a clear need for more postnatal support.

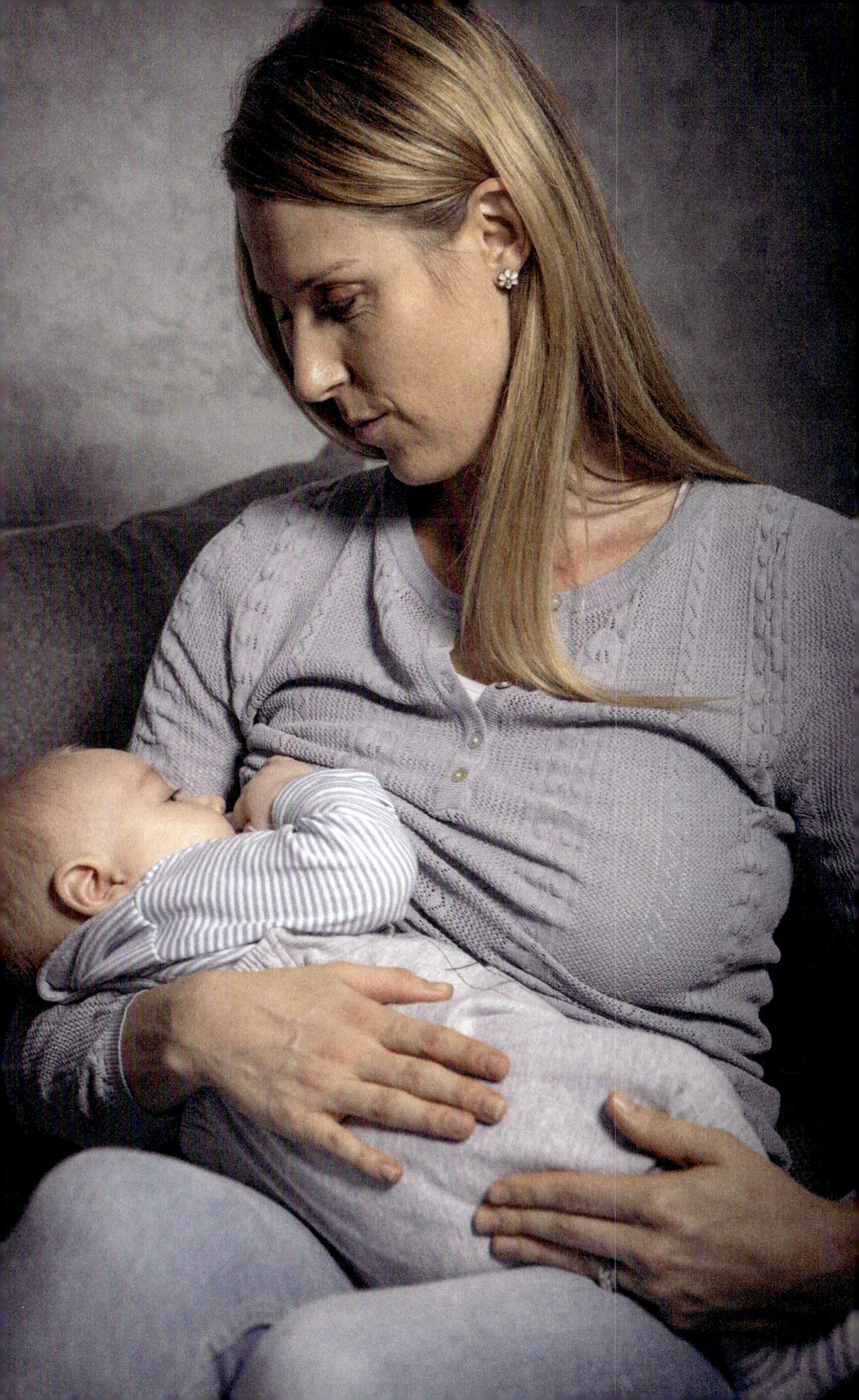

Getting prepared

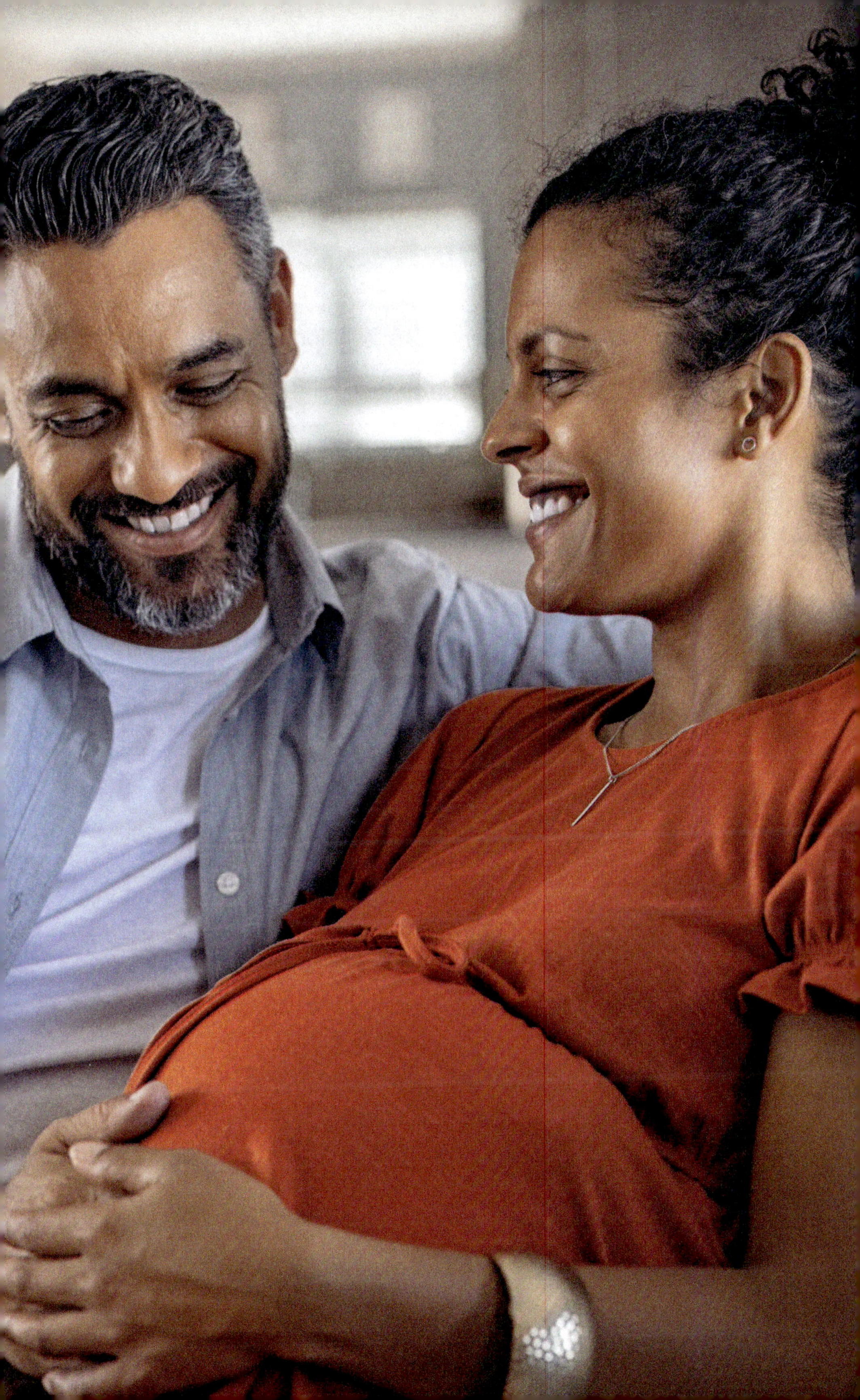

Do I need to prepare my breasts for breastfeeding?
No, you do not.

In some cultures, mothers are encouraged to "toughen up" their breasts in preparation for feeding. Please do not be tempted to do this! Breastfeeding is not meant to cause nipple trauma.

There is no need to use nipple creams or ointments in advance of the birth, and most mothers will not need them afterwards either. Nipples are designed to moisturize themselves, with a natural substance that smells like amniotic fluid to help your baby find their way to the breast. Isn't nature clever?

Shall I express or "harvest" colostrum before the birth?
You can if you want, it is a great skill, but don't be discouraged if it is difficult to get anything out.

From about halfway through pregnancy the breasts start producing a small amount of the first milk, called colostrum. About half of pregnant mothers will notice it – sometimes thick and yellow and sometimes clear and watery. This is the special milk that your baby will receive when they are born and is designed to be rich and easy for little ones to access while they figure out how to suck, swallow and breathe. Colostrum is often called liquid gold due to its colour and amazing properties. It is full of antibodies that start to protect babies from their first feed.

You might be encouraged to get some colostrum out by hand before your baby arrives, and to store it in the freezer for your little one. This is known as colostrum harvesting. Midwives know that hand expressing colostrum is a useful skill for some women after the birth, so it can be good to practise it beforehand. It is particularly recommended for diabetic mothers, and sometimes for those expecting twins or early babies (if near term). Women who succeed in hand expressing may feel more confident. Conversely, some women never see any colostrum and then panic that they will not have any milk (not true!)

Many women do not like handling their breasts, particularly those who have been subject to abuse. Remember that it is not essential to hand express milk before the birth.

Anything that the woman expresses will be replaced by the body, please do not worry that it gets used up. We talk about the process for hand expressing in the chapter on expressing and storing milk.

I am hoping to breastfeed. What do I need to buy?

Breastfeeding parents do not need anything other than functioning breasts and a baby that can drink the milk! This makes most products unnecessary.

Parents often ask whether they should buy items "just in case." This is up to you. Luckily, most items can be purchased relatively quickly and easily should you require them.

If you want to buy something, I suggest:

- A feeding bottle with a newborn/slow flow teat.
- One or two containers of ready-made number one formula milk.
- Breast pads.
- Some nipple shields if you have flat or inverted nipples.
- An electric breast pump if you feel certain that you want or need to express milk.

My suggestion of **emergency bottles and formula** is not intended to undermine your confidence; there may be times when a bottle feed can make all the difference while you work on getting your baby breastfeeding. For more information on types of bottles and formula, see later chapters. Some brands of bottles do not need a separate steriliser if you have a microwave. Most bottles can be sterilised in a pan of boiling water, so you will not need a sterilising machine immediately (or at all).

Many mothers with flat or inverted nipples find **nipple shields** invaluable in the early days. Flat nipples are ones without any protruding part. They are common and not generally do not cause long term feeding issues. Some babies just take some encouragement to learn how to suck themselves into place. Inverted nipples have "dimples" that can look and feel like they are tied into the woman's body. Again, these are normal but can make comfortable feeding more challenging in the very early days. There are many brands of shield, and from my experience Medela size medium, and MAM size 2, suit most women and babies. You can read more in the section about using shields.

Most women find that they need **breast pads** inside their bra to catch drips. Women vary enormously on whether they leak milk between and during feeds. Many need pads from about day three until their supply has settled in a few weeks or months. Pads come in washable or disposable options, and need changing once they feel wet.

Sore nipples are common but rarely necessary! In most cases, breastfeeding should not hurt. For more on sore nipples and how to solve them, see later chapters. Please seek out skilled support to deal with nipple pain. I would not advise investing in silver cups, breast shells, nipple creams or any of the other bits and pieces in advance.

Many mothers know that they would like to, or will need to, express breastmilk. It can be helpful to buy or borrow an **expressing machine** in advance. We look at the pros and cons of different expressing options in later chapters.

Alongside the machine, you might like some **breastmilk storage bags** that are useful for fridge or freezer.

If you are expecting a very early baby, or multiple babies, perhaps consider hiring a hospital grade breast pump. Local pharmacies may offer this service, or look at the Medela or Ardo websites.

Will my baby need extra Vitamin D?

The NHS suggests that babies who are being breastfed should be given a daily vitamin D supplement from birth, even if you're taking a supplement containing vitamin D.

However, babies who have more than 500ml (about a pint) of infant **formula** a day should **not** be given vitamin supplements. This is because formula is fortified with vitamins A, C and D and other nutrients.

The UK NHS recommends that all children aged six months to five years are given vitamin supplements containing vitamins A, C and D every day. Talk to a pharmacist about which supplement would be most suitable for your child, because some also contain other vitamins or ingredients.

Where can I get baby vitamin drops?

In the UK your health visitor can give you advice on vitamin drops and tell you where to get them.

You're entitled to free vitamin drops if you qualify for Healthy Start.

I don't want to breastfeed, what do I need to do? Do I need to inform anyone?

It is your decision to bottle feed, and you do not have to tell anyone.

If it helps, you might like to have a feeding plan documented in your hospital notes, especially if you do not want to explain your decision to every health professional you meet.

As most of our UK hospitals follow UNICEF Guidelines, they tend to assume that a mother will breastfeed, and you might still have to explain your situation to hospital staff. They do not need to know the details, just be clear that you intend to bottle feed and they are trained to assist you to do that safely.

Do I have to take medication to dry up the milk?

Medication to dry up milk is not recommended. Over the counter pain relief might help with the discomfort and you might like to take some milk out by hand if you get sore.

If a woman with functioning breasts does not breastfeed and starts with formula straight away, she will still make milk. This can leave her feeling uncomfortable for several days.

Breasts work by supply and demand. If there is no demand for the milk the receptors in the breast tissue get the message not to make more. However, while this happens a woman might feel full and sore.

It is no longer recommended to take medicines to "dry up" the milk, instead Ibuprofen, paracetamol and cold compresses can help with discomfort or pain. Some women need to express a little milk to remain comfortable and avoid engorgement and/or mastitis. If a woman is unable or unwilling to do this, she will need to look out for signs of mastitis and get antibiotics if necessary.

Decongestants such as Sudafed from the chemist have been shown to reduce breastmilk production, and some people suggest them as a good option for women who do not wish to breastfeed. Please talk to your pharmacist about this, and always take drugs and medicines according to packet instructions.

I am planning to bottle feed with formula. What do I need to buy?

Get some bottles and formula in advance and see which your baby prefers.

If you are planning to bottle feed, at a minimum you will need to invest in:

- Feeding bottles and teats labelled suitable for newborn babies.
- First formula milk labelled number one.
- A bottle brush.
- Steriliser equipment (for most brands of bottle).

Most hospitals ask you to take formula milk with you for the duration of your hospital stay. As they do not have facilities to help you safely make up formula or clean your bottles, they typically require parents to bring in a box of newborn size, ready to feed bottles of number one formula. These come with their own single use teats.

When buying bottles for home use, I would suggest buying one or two bottles that are suitable for newborns rather than committing to a starter pack. While most babies do get on with most bottles, this is not always the case and having a variety may help. Please see the section on selecting bottles. Infant formula is heavily regulated by law, and cheaper brands are thought to be just as nutritious for your baby as the more expensive ones. We cover this in the section on **How to choose infant formula**.

You will need to decide how many bottles to buy based on how frequently you wish to wash up each day. Many sterilisers hold six to eight bottles, so that might be a consideration.

To begin with, you will need a formula labelled number one, as this is suitable for newborn babies. Please do not be tempted by specialist formulas such as lactose free, soy or "hungry baby" formula. These products are not as good for your baby. In a tiny number of cases your GP or paediatrician will prescribe specialist formula if it is necessary.

Everything that encounters milk will need to be washed up well, or put in the dishwasher at 65°C, and then sterilised.

Many parents find that a bottle preparation machine is an essential piece of kit, although there have been recent concerns that these machines do not thoroughly kill bacteria in the milk powder. Bottle warmers are rarely necessary. Please see the section on preparing formula.

I want to express milk and bottle feed my baby. What do I need?

Think about which kit you might prefer, there is a section on expressing later in this guide.

You will probably need:

- Feeding syringes for colostrum or a clean teaspoon or feeding cup.
- A breast pump.
- Milk storage bags.
- Bottle feeding equipment; bottles and teats for newborns plus a means of cleaning and sterilising them.

In the first few days after birth, you will need to express milk using your fingers and collect it in a feeding syringe, clean teaspoon or any clean collection container. If you are in hospital, these things can be provided, otherwise you can find them at a pharmacy or online.

You can use a breast pump after you are making about 5 ml or more each side. Most parents buy, borrow or rent a breast pump. Please see the section on expressing and storing milk.

Expressed breastmilk can be stored in the bottle in which it was collected or poured into special breastmilk storage bags for the fridge or freezer.

You will also need bottles and means to clean and sterilise them, please read the chapters on bottle feeding.

Your baby's first breastfeeds

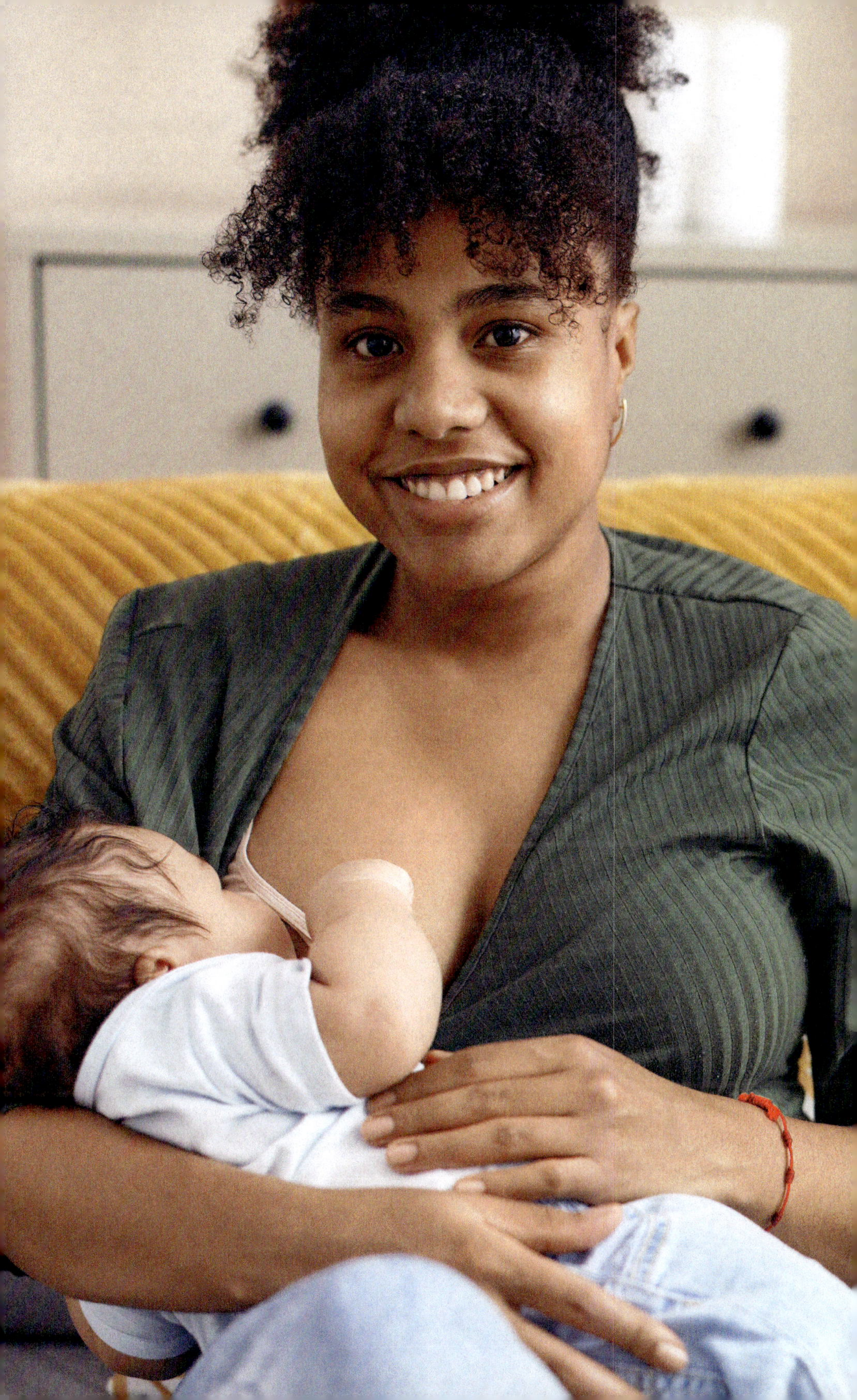

Breastfeeding is the perfect start to a close and loving relationship between you and your baby and has lots of benefits for you both.

Breastmilk provides your baby with everything they need nutritionally for the first six months of their life and remains the main source of nutrients and protection for at least the first year and beyond.

It can take a couple of weeks for you to become confident in your breastfeeding. This is completely natural. Trust in your body's ability to continue to nurture your wonderful baby, and enlist as much help and support as you need. Some parents feel embarrassed about seeking some help, but it can make all the difference to both your comfort and confidence in the early days.

How does breastfeeding work?

Breastmilk is made in clusters of cells in the alveoli of the breast, and milk ducts carry it down towards several small holes in the nipple. Milk builds up in the breast until it is needed, when the hormone oxytocin makes it come down the milk ducts to feed your baby. Your body knows how much milk your baby drinks and regulates supply to match the demand.

You have probably noticed breast changes through your pregnancy.

The first milk, **colostrum**, is produced from mid-pregnancy and about half of expectant mothers will see it. It ranges from bright yellow to white or clear. This milk is perfect for your baby immediately after birth. It will provide all the food and water that your baby needs, plus vital immunity and protection from harmful bacteria and viruses.

No matter how your baby is born, the delivery of the placenta stimulates a hormonal cascade over the days and weeks ahead. Initially babies feed on increasing amounts of colostrum, and then after two to five days the milk volume goes up and some mothers will notice their breasts become fuller between feeds. This is often referred to as the "milk coming in" but remember that your baby was fed on colostrum before then, they were not starving!

Two principal hormones are involved in breastfeeding - prolactin and oxytocin. Prolactin stimulates milk production: you can boost it with plenty of skin to skin contact with your baby, and lots of feeding (or expressing) at the breast. Oxytocin produces the "let down" of milk as your baby, determining how quickly the milk flows and the length of feeds.

Breastfeeding works primarily by **supply and demand**. As the milk is taken out of the breasts the alveoli replace it for the next feed. Your body will know how many babies you have, and how hungry they are, by the supply and demand cycle.

Every woman and baby combination is unique with a complex interaction of different physical, practical and emotional factors. As you get to know your body and your baby, you will figure out what works best for you both.

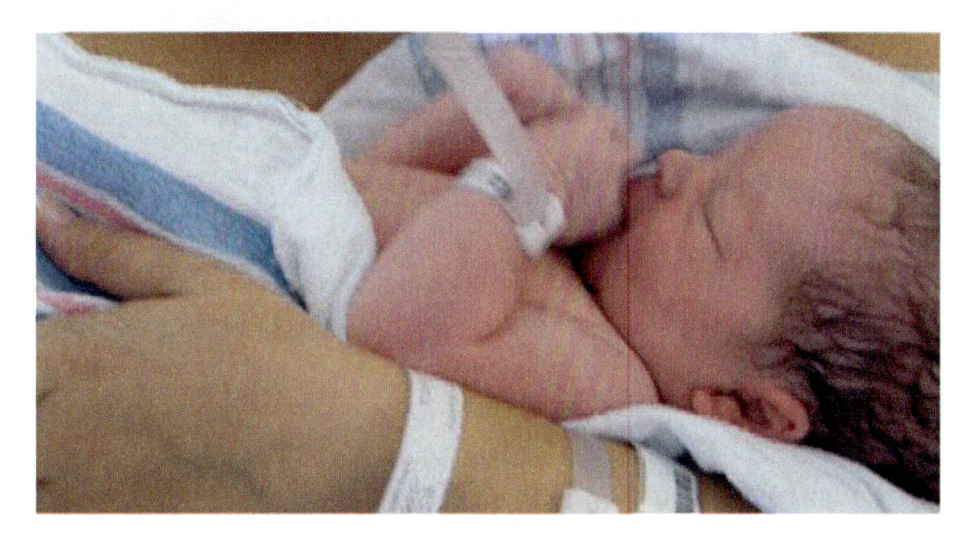

How do I start breastfeeding my newborn?

As soon as you can, snuggle down with your baby on your chest, skin to skin. Some babies are hungry and want to suck the instant they are born; others might take an hour or two of getting to know you before latching on.

We know that skin to skin contact fills mothers and babies with wonderful happy hormones, reduces stress and promotes healing and bonding. Babies respond well to hearing the heartbeat and voices they heard in the womb. They do not need to waste precious energy in keeping warm, and benefit from their mother's friendly bacteria.

Some babies are born wide-eyed and alert, keen to feed. Others are much sleepier, especially after a mother has had pethidine, an epidural, instrumental or caesarean birth. Try to encourage your baby to feed within the first couple of hours. If your baby will not feed, hand express milk for them. Midwives in the delivery and recovery areas often have more time to help you than those on the postnatal wards. If your baby has not fed in the first six hours, get some support in breastfeeding your baby and feed hand expressed colostrum. There is more on expressing later in this guide.

Keep yourself comfortable by reclining gently back on a bed or chair, using pillows to support your head and back as needed. Make a nest for yourself. If you have had a caesarean birth, you might be more comfortable with your baby across you and away from your bump and scar, or putting a pillow over any areas that are sore.

Your baby will use natural instincts to find the nipple. When they have their tummy against you, they will use their arms and legs to crawl into position. They feel your breast tissue with their face and hands, they are looking for a darker areola and they are smelling for the nipple. When they find the nipple they will bob their head and open their mouth to latch on. Gravity holds your baby in place and helps them to get a large mouthful of breast.

Some babies need some gentle guidance and a bit of patience in the early days. Left alone to get on with it, mothers and babies do a natural "dance," the mother supporting her little one to get comfortable. Some women naturally shape their breasts to make it easier for their baby.

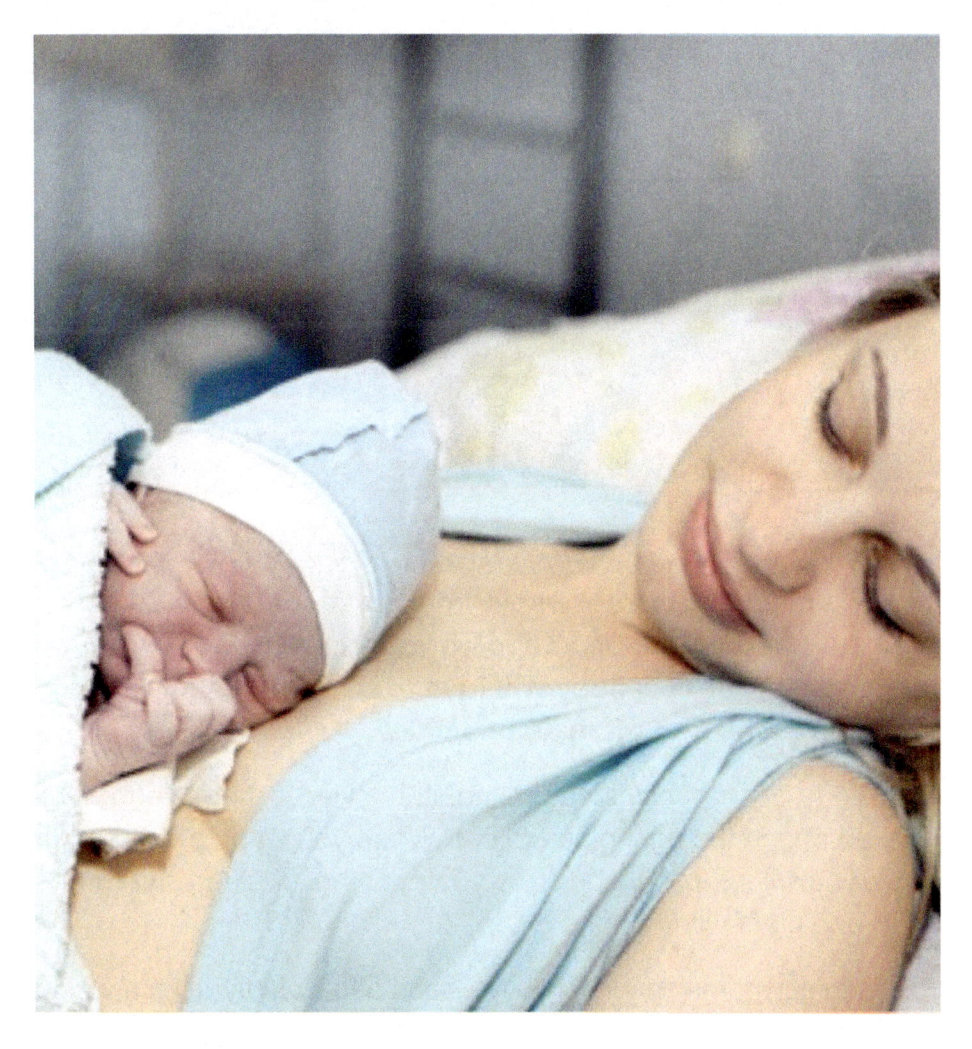

These laid-back feeding positions are very natural and work well for all feeds. You will do well to lie around with your baby on your body for as much time as you can. Your baby will naturally rouse from sleep, have a little feed, then sleep some more.

If immediate skin to skin contact is not possible then Dad or a birth partner can do it until you feel ready. While it is wonderful to hold your baby skin to skin straight away, it is never too late, and can be lovely even weeks or months after the birth.

When do I get my newborn baby dressed?

You will need to get your baby dressed at some point when they are not going to be snuggled with their grownups. There is no rush! Your new baby will love to be skin to skin with you for as long and often as possible.

Your body is your baby's natural habitat, it keeps them warm, relaxed and in the right place to eat. When you want to dress your baby, you will probably choose a vest and then a sleep suit (an all-in-one outfit with buttons or zip).

Full term newborns generally do not need a hat when they are inside a building apart from when they are in a cold operating theatre.

A good general rule is that your baby wears one more layer than you. Check your baby's core temperature by putting your hand on their front or back inside their vest.

Parents often worry that their baby will scratch their face with their surprisingly sharp little nails. It is tempting to dress them in scratch mittens or to fold over the sleeves of their sleep suit, but babies need their hands free when they are breastfeeding. Touching the breast is an integral part of how they orientate themselves and latch on. It is lovely to breast and bottle feed your baby skin to skin or lightly dressed as this works with their natural feeding cues.

Positioning and attachment

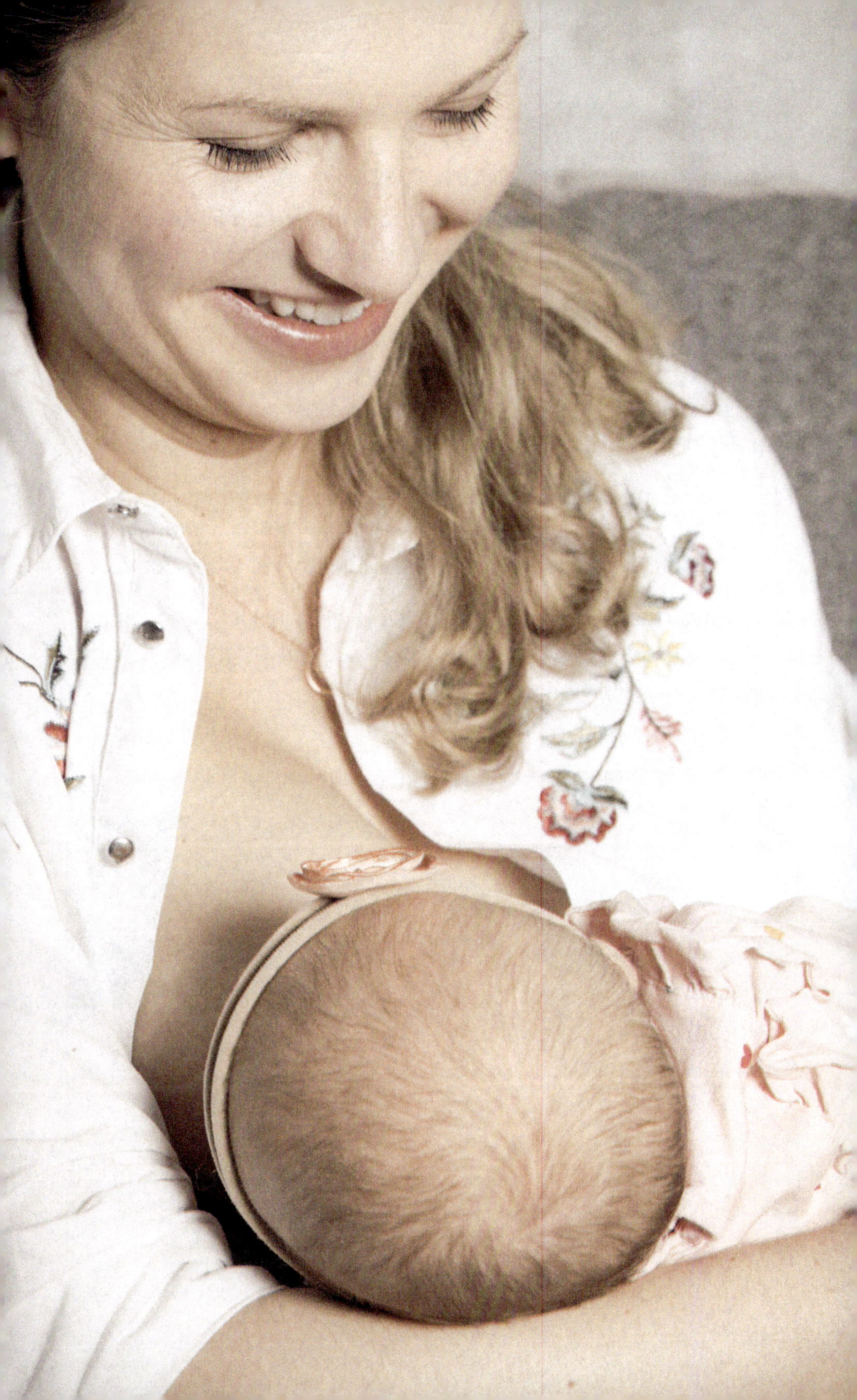

How do I feed my baby comfortably? Which feeding positions work well?

There are lots of breastfeeding positions, and it is good to experiment and find the ones that work best for you in different locations.

There are a few things that the positions have in common. They need to be comfortable and sustainable for you and your and baby. You should not need to move your breast away from where it naturally hangs. Your baby will have their tummy against your body.

You need to find a place where you can relax pain free for up to an hour. This might mean reclining on pillows, lying on your side in bed, sitting up or relaxing back in a chair, whatever works for you.

If you have had a caesarean birth you might wish to cover your wound with a small rolled up blanket or cushion to protect it from being accidentally knocked. Lying on your side won't feel good for a few days.

The following pages detail some of the most popular positions.

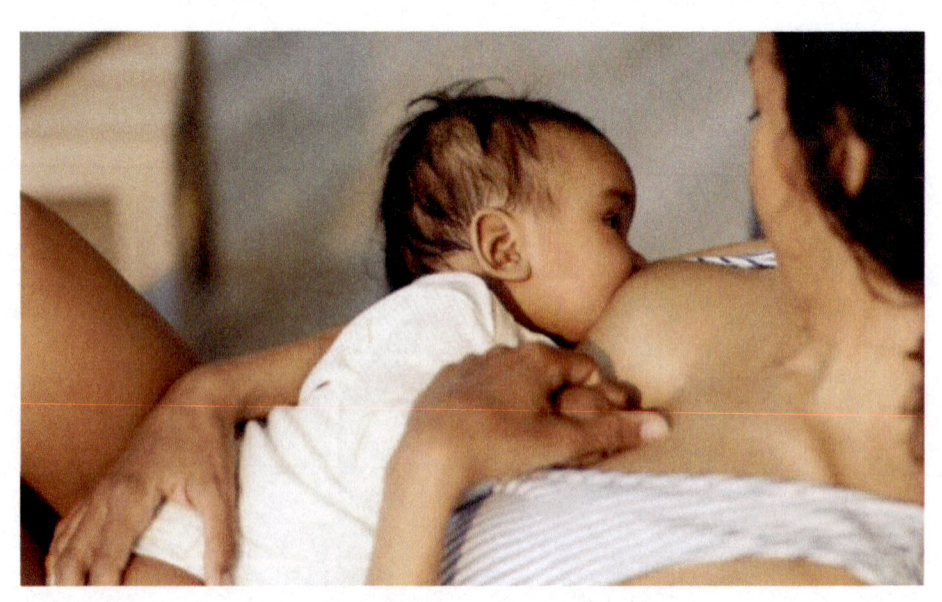

Laid-back feeding

Many women use variations of laid-back feeding positions a lot of the time, especially when relaxing at home. These do not need to be taught, a mother and her baby instinctively know what to do. The baby will usually feel the nipple, open their mouth, drop onto the nipple and start sucking by instinct.

Mothers can help and guide their baby if they want to, it is a beautiful, interactive and loving dance between them.

In the picture opposite you can see that the baby's body is against mum, with their hands either side of the breast. Gravity holds the baby in contact with their mum and encourages a deep latch at the breast.

Cradle hold

We use this term for positions where a baby is cradled with their head towards their mum's elbow. Leaning back slightly, like the mum in the photo, allows your baby's weight to rest naturally on your body. In the picture you can see that the baby has their tummy against mum and can reach the breast. The mum is using a pillow to support her right arm and looks very comfortable.

It is important to leave your breast where it naturally wants to hang. Check whether your baby can reach the nipple without you moving it for them. Your baby can be held horizontally across you or, more likely, with their bottom resting on your lap towards or on your opposite thigh.

Cross cradle hold

The cross cradle hold gives a woman more control over her baby's position. The mother supports her baby with an arm along the baby's back and hand behind their shoulders and neck.

Babies don't mind if you hold their neck and shoulders, but they do not like being pushed by the back of their head. This is especially true if they are sore or bruised from a long labour or instrumental birth. Pushing your baby's head also tends to result in a nosedive position at the breast. It is more comfortable for your baby to tilt their head back slightly as they feed, and we never hold a baby's head against the breast.

There is nothing stopping a mother from starting in one position and then rearranging herself to feel more comfortable.

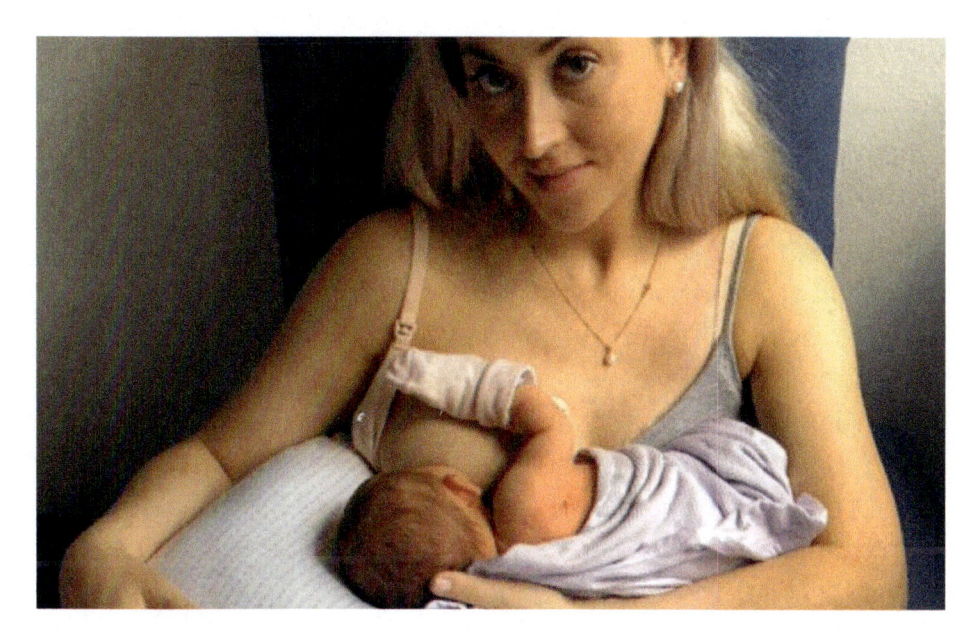

The straddle/koala hold

Many mothers and babies enjoy this "koala" or straddle hold, where the baby has their legs either side of their mum's thighs.

You can hold your baby like the mother in the image or curl a supporting arm around your baby's body, so they rest their head on your upper arm. This can allow a free arm for holding your drink, phone, or cuddling someone on your free side.

The koala hold is tricky for mothers with pendulous breasts and nipples that point downwards.

Feeding on your side

You might find it lovely to feed your baby while you lie on your side in bed. This is often super comfy for mothers with a sore perineum, stitches, haemorrhoids, or just those who fancy a rest!

All you need to do is lie comfortably in a flat bed, with one arm up to support you and your legs slightly bent. We call this the cuddle curl. Your baby will lie on their side with their tummy against yours. They should be able to reach the nipple easily with their head slightly tipped back. If they nosedive into you, try moving them slightly down the bed and snuggle their body and bottom in closer to your body. This should encourage them to tilt their head back slightly and free up their nose.

The side lying position often does not feel great for the first few days after a caesarean birth, when the wound can feel tugged. You will know when your body is ready to lie on your side.

Mothers also often worry that they will fall asleep in this position, but this is OK if common sense safety precautions are in place. See the section on feeding your baby safely in bed.

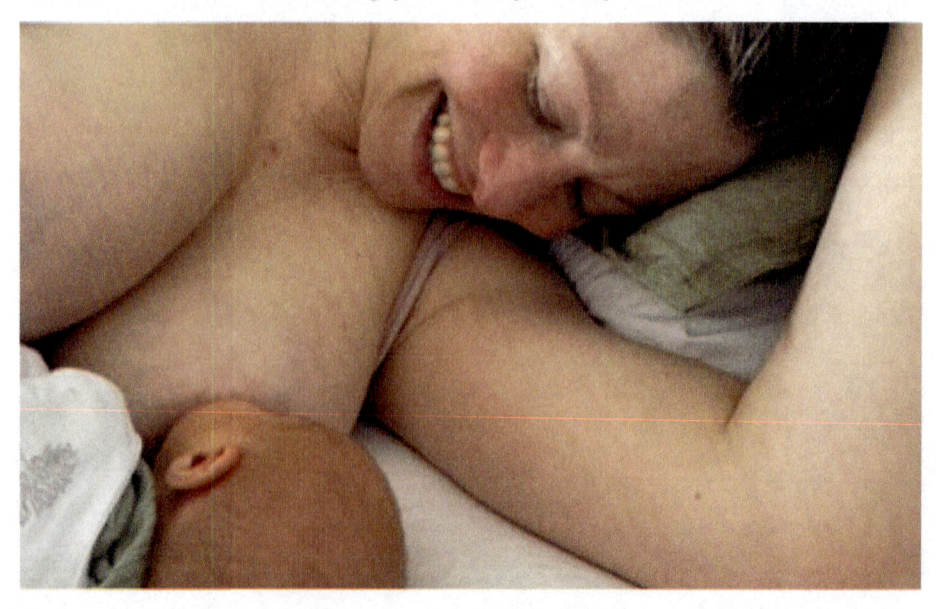

Using feeding pillows

Pillows can be helpful for some mums, allowing them to sustain a feeding position by having their arms or baby's weight supported. They can be super useful for mothers of twins, those feeding underarm, and for keeping a baby away from a sore tummy and caesarean scar. They also allow some mothers to feed with at least one hand, and possibly both hands free.

There are many breastfeeding pillows available, and some pregnancy pillows are advertised as being suitable for feeding. I find that ordinary medium to firm bed pillows work just as well. Some of the U-shaped pillows and those with rounded tops sometimes makes life harder, as the baby seems to fall into the gap between the pillow and their mum's body.

If you fancy using pillows to support your feeding position, make sure that they support your baby's weight without you needing to hunch or round your shoulders. Try tucking them in close to your body and make yourself a "breast shelf" for the breast and your baby to lie on. The pillows should keep your baby in a perfect position even if you let go, so that you can relax your shoulders and get on with other things like eating your dinner or using your computer.

Alternatively, get yourself and your baby comfy and bring in pillows to help you to sustain your position. For example, some women like pillows to support their arms or elbows and can then feed one handed.

The underarm or rugby/football hold

This position can work well for mothers in the early days of breastfeeding, after a caesarean section, mothers of twins, and those with larger breasts where the nipple points downwards or to the side.

Your baby will lie on their side, under your arm with their feet pointing towards your back. As you will see, there is often not much room for your baby's body to stretch out when feeding this way. You might need to sit slightly towards the front of the chair, bed or sofa. Put pillows behind you to support your back but still allow space for your baby's legs. You might like an additional pillow to support your baby's body.

A word of caution here, please don't let the weight of your breast rest on your baby's chest. Your baby needs to lie on their side.

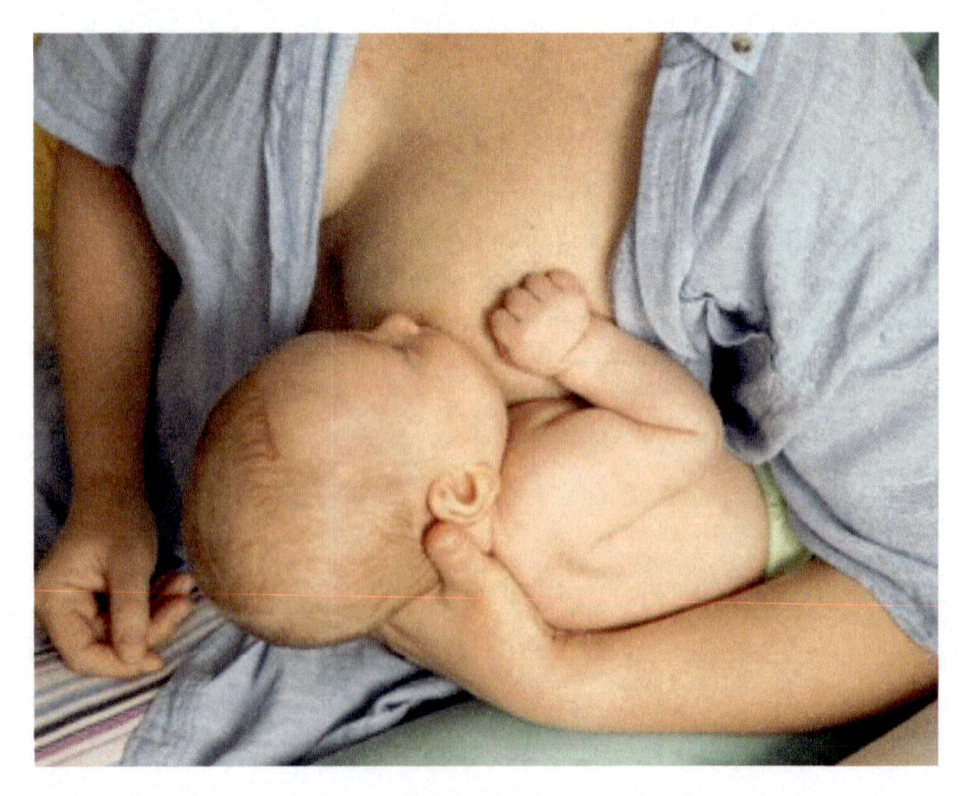

How do I keep my baby comfortable?
Your baby needs to be:

- Tummy to tummy with you.
- Able to reach the nipple without you moving it to them.
- Able to move their head freely.
- In alignment, so their neck isn't twisted.
- Able to drink with their chin and cheeks touching your breast.
- Touching something with their feet.

Your baby needs to be comfortable so that they can latch on and feed. Whatever position you choose, your baby should be **"tummy to mummy,"** facing you without needing to turn their head. The most natural positions have your baby well supported on your body, so that they can use their instincts to feel the nipple, open their mouth and start reaching for the breast. Laid-back positions like this also avoid needing to push on a baby who is sore from a long labour or instrumental birth.

Babies often feel more secure when their **feet are touching** you or the environment, rather than dangling.

Feeding skin to **skin or lightly dressed** is awesome when possible. If you are dressed, I suggest that you move your bra and other clothes away from the breast, so that your baby has plenty of space to sense you and latch on. It is easier for your baby to use their instincts if they have their arms and hands free, rather than being wrapped or wearing mittens.

Once you have both got comfy, look at which way your nipple is pointing. This will vary depending on how you are sitting or lying. **The breast always stays where it belongs; you should move your baby to your breast.**

It is tempting to pick your breast up and give it to your baby, but then your baby is going to be holding your nipple somewhere it does not want to be. This can kink your milk ducts, get your nipple squished, cause pain and make life much harder for everyone! Sometimes babies end up with blisters on their lips from holding tightly to a breast that wants to pop back out of their mouth.

Mothers with nipples that point downwards might find some breast support helpful. This can be something like a rolled-up muslin under the breast or using feeding pillows under the breast to provide support.

If gravity is working for you, your baby will lift their head and latch on by instinct. This can take a few minutes while you are learning, and it can seem like your baby is getting frustrated. You can help as much as you like, it is a joint effort. If your baby is screaming you might invite them to suck on your clean finger, so they calm down and remember about sucking, then try again.

Avoid holding your baby's head and forcing it onto the breast, this can make your baby panic and turn away. They need to be able to reach your nipple with their chin on your breast and be able to move their head slightly back if they need to. Gentle support of their head and neck is fine.

How can I encourage my baby to open their mouth wide?

Your baby needs a big mouthful of breast.

Your baby's mouth should be close enough to your nipple for them to reach it easily. You might like to allow the nipple to brush baby's lips, and/or squeeze a little milk out with your fingers to tempt them to open their mouth. When they feel and smell you, their instinct is to root around for the nipple with their mouth open and tongue down.

Most breastfeeding institutions recommend starting with your baby's nose or top lip next to the nipple, so that they are encouraged to tip their head back and open their mouth wide. Your baby should still be able to reach the nipple, it is frustrating for you both if they open their mouth but still cannot reach you. I find that when your baby is on top of you and leading the process, you don't have to worry about this quite so much.

If your nipples hang low, then gravity is not working for you and your baby cannot latch without help. You will need to give them some support. Hold them close to your body and guide them on with your hand along their back, shoulders and the back of their neck. Your baby would prefer not to have the back of their head pushed.

Your baby will start the feed with some little "fluttery" sucks, and then start to swallow milk.

How do I know it is a good latch?

If your baby is well attached, their cheeks and chin will be touching the breast and not much, if any, of the areola will be visible. Their cheeks will be full and rounded, touching the breast equally each side. You probably won't see their lips because they are so close to you, but they will have a nice wide-open mouth. There will be very little movement of the breast, and certainly no tugging of the nipple away from where it belongs.

If you are in pain after a slow count of five, look at your breast. If you can see the skin tugging towards your baby with each suck, move them slightly towards the tugging and see whether that helps. Tuck their body in nice and close to yours. If it still hurts, take your baby off by inserting a finger in the corner of their mouth and between their gums, and try again. Some sensation is normal though, this is new to both of you!

Here is a picture of a good latch. The baby is deeply attached with a big open mouth and has rounded cheeks. Their head is free. Mum's nipple has not been moved from where it naturally belongs, her upper arm gently supports her baby's head. The baby has full rounded cheeks touching the breast so we cannot see much of the areola. Babies often feed with their eyes closed.

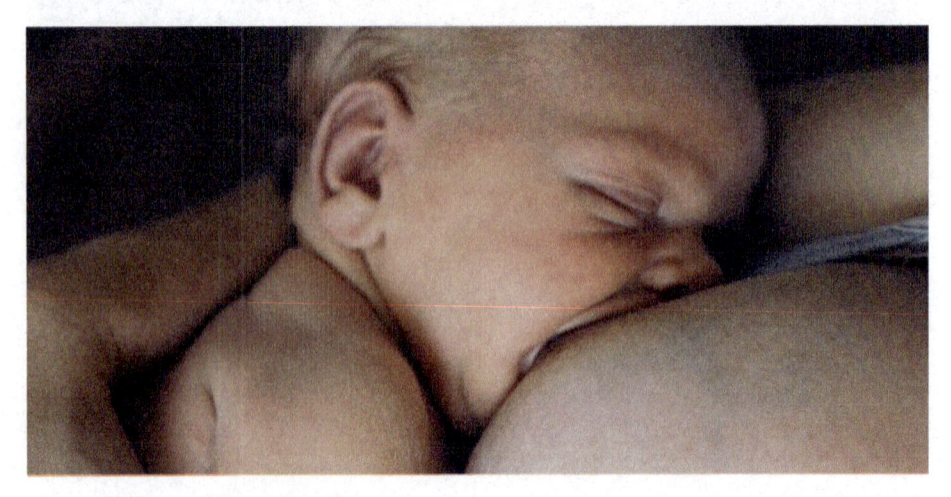

In contrast, the following picture shows a **shallow, uncomfortable latch**.

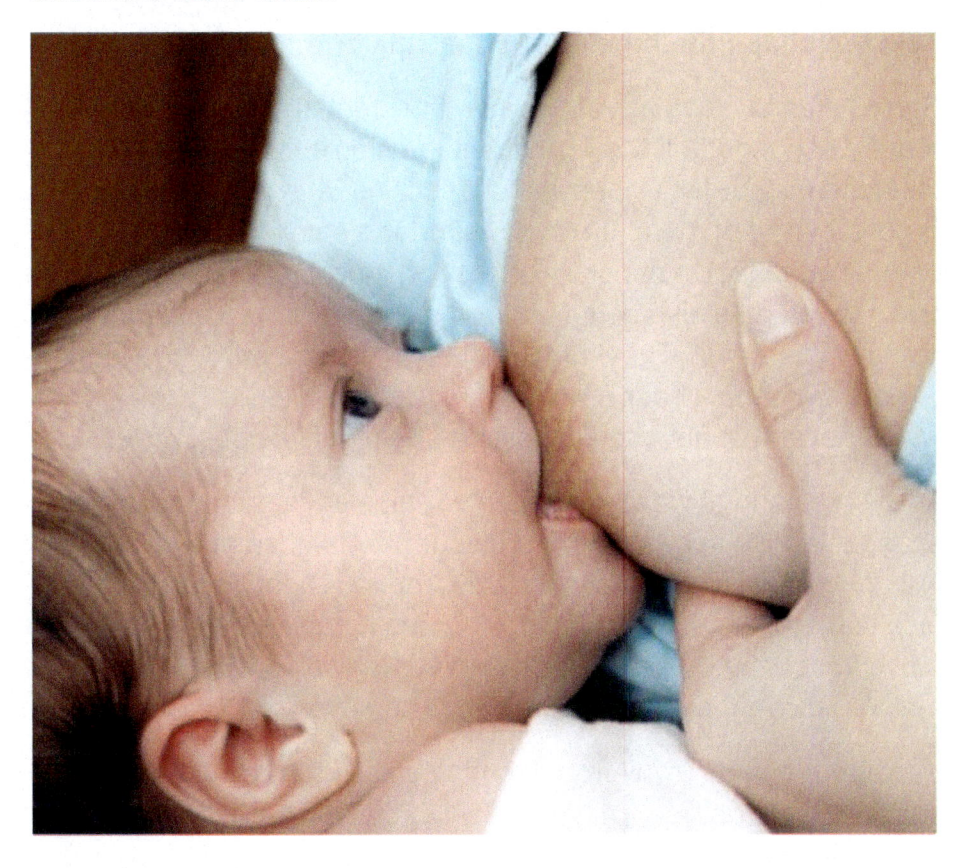

The baby's lips are visible, they are not wide or turned outward. The baby's cheeks are not touching the breast tissue. The baby is hanging below the breast, pulling it downwards.

Although many sources suggest that the mother holds her breast at the start of the feed, this baby's mother is squishing her breast in a way that is making it harder for her baby to latch on. It would be more effective to use her thumb and fingers on the breast above baby's nose and below their chin. This could make the nipple the shape of a cookie from the baby's perspective. Sometimes it helps to get another adult or feeding specialist to watch what you are doing.

Do I need to squish or shape my nipple?

Most babies can latch on by themselves, but sometimes it helps to shape the nipple for your little one.

Babies start sucking by reflex when they feel the nipple on the roof of their mouth. If they don't feel the nipple in the right place they often cry or fall asleep without sucking, even when the nipple is right in front of them. It can be hard for a small baby to latch onto a nipple that is flat or inverted, or if mum's breasts are firm and swollen. It is also common for a new baby not to open their mouth wide at first. They might be tiny, or have a sore face, head or neck from the labour and birth process. This tends to improve with time, and shaping the nipple can help until it does.

There are various ways to shape your nipple, just remember to squeeze above and below where your baby's nose and bottom lip will go (otherwise you make it harder for your baby to latch!) It is also important to shape the nipple where it naturally hangs, and move your baby to the nipple, not the other way around. When mothers move the nipple to the baby, they often get sore nipples.

Keep your fingers away from the nipple itself to allow your baby space to latch on.

Once your baby is sucking happily, please release the shaping fingers and relax.

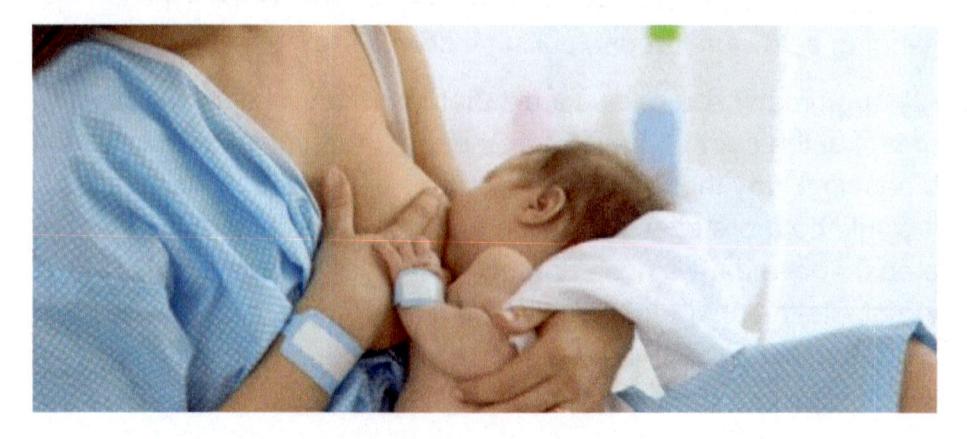

Why isn't the "nose to nipple" thing working for me?

You might need to try a different technique that suits you both better. Laid-back feeding works well with your baby's natural instincts.

The NHS and many other organisations teach a standard breastfeeding technique that works well for many mothers and babies. It is led by the mother and not the baby.

The idea is to sit up, and line your baby up with their nose next to your nipple. When they open their mouth wide, you guide them onto the nipple, keeping their chin in contact with the breast and with the nipple aimed at the roof of their mouth. This method encourages your baby to feel the nipple on the roof of their mouth and start sucking. If your baby is well latched and you have no nipple pain, that is brilliant.

The standard "nose to nipple" technique can work for big, robust babies who will open their mouths wide, lean their head back and patiently wait for their mother to scoop them onto the breast leading with their chin. However, I see hundreds of mothers with damaged nipples and hungry babies who cannot "get it."

When I watch a sore mother feed this way, I notice similar things. Mum and baby are fighting against gravity. The mother, tense and anticipating nipple pain, holds her baby away from her body waiting for the "wide open mouth." Her baby may be desperate to feed but cannot feel the breast or reach it, so flails their arms, sucks their fingers, boxes the breast and screams. Alternatively, her baby might not show hunger cues at all.

Very often, a mother will move her nipple towards her baby's mouth so that her baby can reach it. This results in the nipple being bent over in her baby's mouth, causing the mother pain and preventing optimal milk flow. The baby can struggle to stay on, bob off a lot, and sometimes gets pressure blisters on their lips. When her baby comes off the mother's nipple is flattened. The mother will typically spend the entire feed holding the breast towards her baby's mouth, so that it does not ping back to where it naturally wants to hang! She will often have very sore shoulders from bringing her nipples to her little one and hunching forward.

If the "nose to nipple" technique is working for you then brilliant. But if it feels unnatural or is hurting, please try something else. Laid-back breastfeeding almost always helps.

There are lots of ways to breastfeed. Look and feel how you and your little one are most comfortable. Forgive the analogy, but if you were told that there was only one way to have sex and that any other way was totally wrong, you would probably say *"No, I don't think so."*

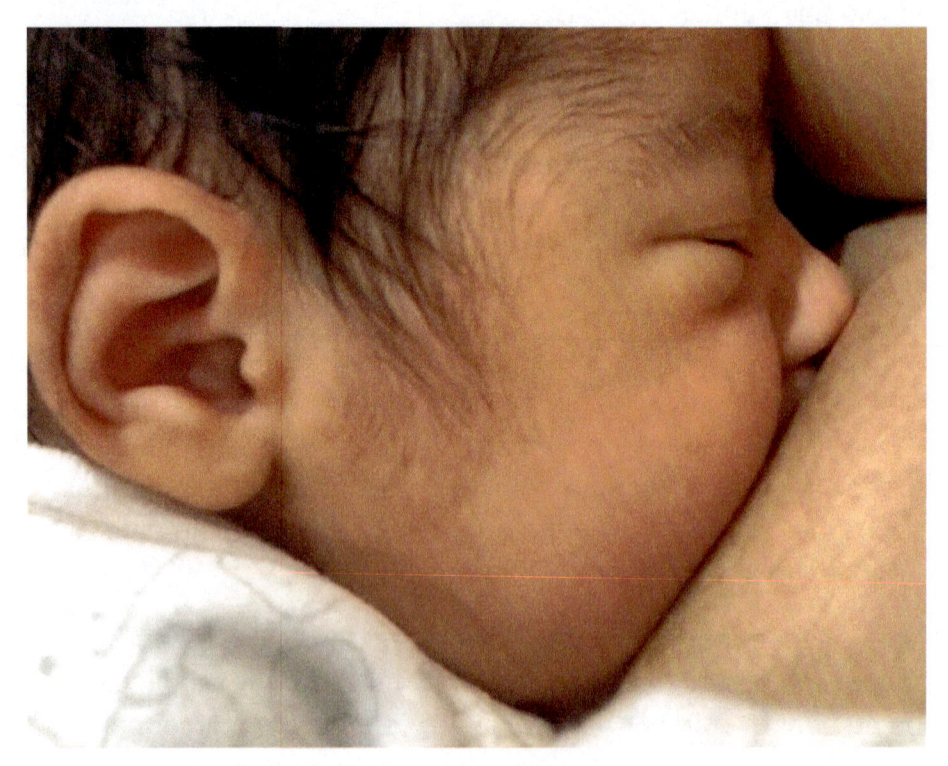

Is it OK for my baby's nose to touch the breast?

Your full term and healthy baby is designed to be close to you and will prioritise breathing over feeding if necessary. Snuggle your baby's body close to yours, as this stops them from nosediving the breast and keeps their nostrils clear.

Many new mothers ask whether their baby can breathe if their nose is touching the breast. If your baby is full term and well, and you are not forcing their head against the breast, they will usually adjust themselves so that they can breathe comfortably. If the end of your baby's nose is touching the breast, then their nostrils are well designed to flare open on each side. Obviously, keep an eye on your baby.

If you can hear your baby snuffling or they look too buried in your breast, experiment with moving your baby slightly towards their own feet and making sure that their body is tucked in close to you. This will often allow their nose to sit slightly further away from the breast. Be led by your comfort, and how your baby is behaving.

Ideally do not hold the breast back from your baby with your finger, as this may result in a shallow latch and risks blocking your milk ducts.

My newborn baby won't latch at all! What can I do?

Please feed your baby by expressing colostrum into syringes or a cup, or if colostrum isn't available use some formula. In the meantime, seek some skilled support.

It can be alarming and frustrating when a baby will not latch onto the breast at all. Some babies are alert, hungry and angry but cannot seem to figure out how to suck at the breast. Others seem disinterested in feeding at all.

This is usually a temporary issue, so we feed the baby and protect your milk supply until it is sorted out.

Please get some help from your midwife to collect colostrum by hand and feed it via syringe or cup into the baby's mouth. There is more about this in the section on expressing milk. Your baby will most likely drink whatever you can make. Aim for 2-5 mls or more per feed on the first day, and this will increase in the days head until you notice much more milk. Formula feeds would need to be more than this, perhaps try 10-30 ml per feed in the first instance and increase it if necessary. You can use a breast pump effectively once you are expressing more than 5 ml from each side.

In many cases, a colostrum feed from a syringe will either calm your hungry baby or wake your sleepy baby so they are more likely to breastfeed. In the meantime, relaxing skin to skin or lightly dressed will be excellent for both of you. I recommend laid-back positions where possible, as they promote your baby's natural instincts to find the nipple and start to suck.

If your baby is desperate to find the nipple but cannot latch, you might like to shape the nipple before they latch on to make it easier for them to find and feel the nipple in their mouth. Hold the shape while they start to suck and then gently let go. You can help to remind them about sucking by giving them a clean finger to suck first. If shaping by hand does not work, you might try using a nipple shield. Please be aware that not much colostrum gets through a shield, they are most effective when the milk comes in. Using shields in the first couple of days keeps something lovely happening at the breast, but your baby might also require more milk by syringe, cup or bottle.

In the meantime, keep up the cuddles skin to skin or lightly dressed. You will get there in time.

Why won't my baby latch on?

Hungry babies might be trying desperately to find the nipple, but can't latch on for some reason for example:

- Being held in a suboptimal position.

- Pain from birth injuries such as bruising, a stiff neck or tight jaw.

- They have an anatomical issue like a tongue-tie.

- The nipple might be an unhelpful shape or size for the baby right now (severely inverted or flat, physically too big for baby's mouth).

- The breast could be engorged and tight, so it is hard for the baby to draw the nipple into their mouth.

Sleepy or disinterested babies might not latch because:

- They went through a long labour or complicated birth and they are tired and/or bruised.

- They have low blood sugar.

- They are jaundiced, which makes them more sleepy.

- Their tummies are full of mucus, meaning that they are not feeling hungry.

Feeding quantity, duration and frequency

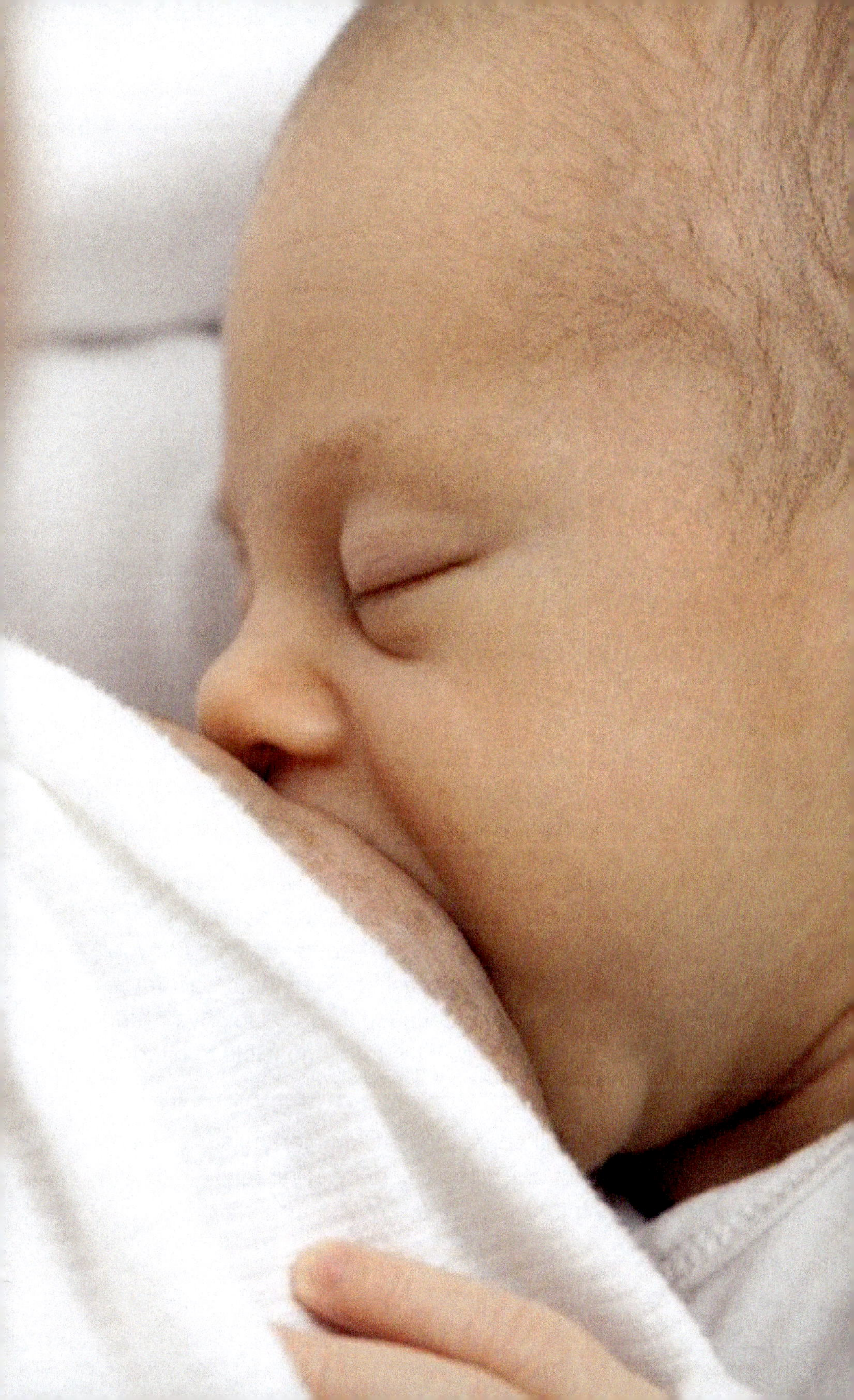

How often should I feed?

On the first day, you can expect your baby to feed at least 5-6 times. After that, babies should be feeding about 8-12 times in 24 hours.

Your baby will not space these feeds out equally, and you might feel like they are clustering feeds so that they roll into each other. Most parents can trust the process and be led by their baby.

How can I tell when my baby is hungry?

As a rule of thumb, if your baby is waking after a sleep, they will be hungry. When your baby is ready to feed, they will also show some signs or "cues," such as licking their lips, turning their head, trying to suck their hands and searching for a nipple with their mouth open!

Try to offer your baby the breast or bottle before they start to cry. Some babies cry quickly, so don't worry if this is not always possible. If your baby is really agitated, it might be necessary to calm them before offering the breast or bottle. Try putting them skin to skin on your chest and talking gently, stroking or shushing them, and/or getting them to suck on your finger.

As your baby receives both food and comfort at the breast, it is always appropriate to offer the breast if they are searching for it. This is part of being responsive to your baby and helps baby to learn to trust you. Your baby might ask to go to the breast for a cuddle, for help going to sleep, because they feel lonely or feel under the weather, that is all fine. It is rarely possible to overfeed or "spoil" a breastfed baby.

A bottle fed baby will still appreciate sucking for the same reasons, so you can offer them a clean finger or pacifier to calm them without feeding them if they are not hungry.

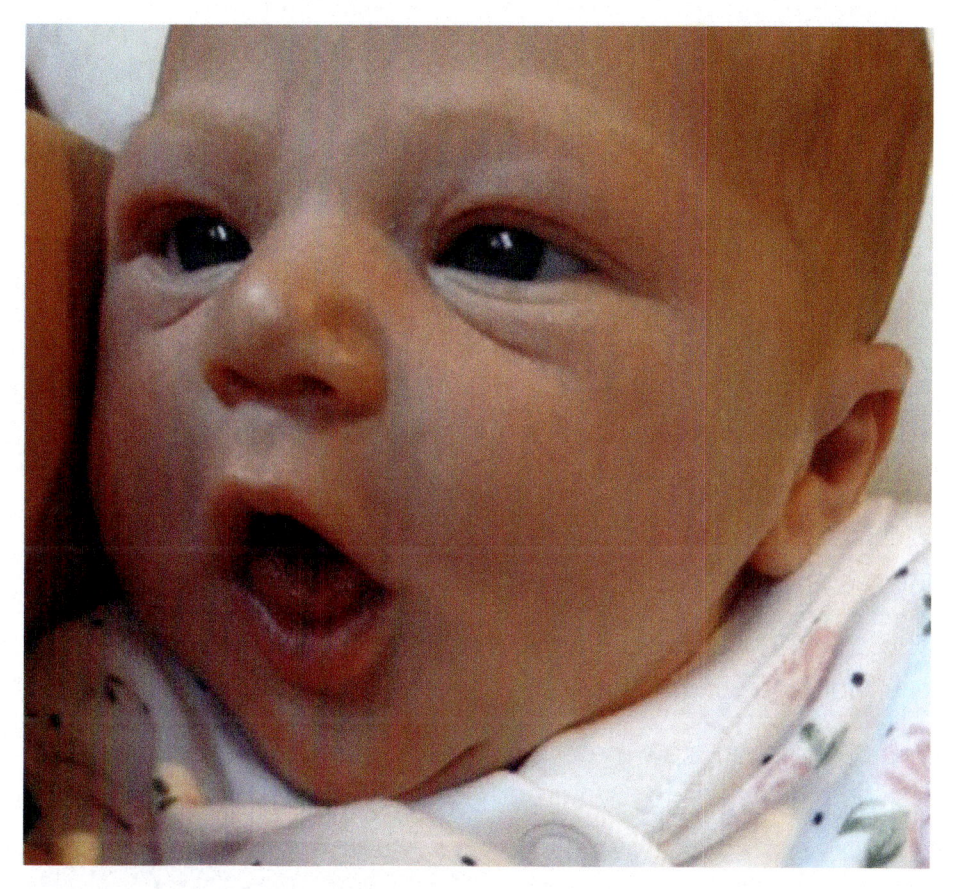

How much breastmilk does my newborn baby need?

Newborn babies do not drink large volumes in the first day or two after birth. Feeds are small and the volume increases steadily through the first week.

Your body provides your baby with small amounts of concentrated, protective colostrum. This is great, because it can take a while for babies to get the hang of the suck-swallow-breathe pattern that they need to feed. Full term babies have fat stores, like their own packed lunch, and carry extra water that they need to wee out. As they get capable of taking more milk, your body will provide it.

Remember, the feed volume will vary from mum to mum and baby to baby, these illustrations are a very rough guide.

- Day one: 2-10 ml per feed.
- Day two: 5-15 ml per feed.
- Day 3: 15-30 ml per feed.
- Day 4: 30-45 ml per feed.
- Day 5: 45-60 ml per feed.

After the first few days, paediatricians would say that babies need 150-200 ml milk per kilo of birth weight in 24 hours.

So, a 3 kg baby would need between 3 x 150 ml = 450 ml and 3x 200 ml = 600 ml in 24 hours. Divide the total by eight feeds and the range is 56 – 75 ml per feed.

The amount of milk does not keep going up exponentially! From one month to six months, babies take on average 750 ml a day.

You don't have a flow meter on your breasts! So, you will have to rely on other signs that your baby is feeding well.

My baby is latched! How do I know they are drinking?

Look for your baby's jaw movements. A swallow cycle takes about a second. When their mouth is full of milk their jaw lowers and there is a tiny pause. Then you might hear them gulp and make a little "cuh" noise as they swallow.

When your baby first comes to the breast they will suck rapidly, like a fluttery movement. This is your baby asking for the milk to let down. After a few seconds the milk starts to flow and your baby may settle into more rhythmic sucks, but with some pauses.

When your baby is swallowing their jaw will move down lower and there will be a tiny pause, you might also hear a gulp or a little "huff" from your baby's nose. When your baby is sucking and swallowing milk, the rhythm will typically be one suck per second. They will suck and swallow for a while, pause for a bit, and then start sucking again.

In the early days your baby might take several sucks before a swallow, as colostrum is low in volume.

Towards the end of a feed on each side your baby receives less milk, but it has a higher fat content. This is called hind milk. It helps your baby to feel full. Your baby might suck several times before eventually swallowing. Sometimes your baby might do fluttery rapid sucks later in a feed, to stimulate further let downs.

When should I switch sides?

Most babies benefit from being offered both breasts each feed, swapping when the baby seems to be falling asleep on the first side. Some mothers might switch between breasts several times in one feeding session. Others have so much milk in each breast that they rarely need to use both per feed. Babies are good at taking what they need.

"I was told that my baby must spend at least 20 minutes on the first breast so that he gets the hindmilk/pudding milk. Is this true?"

They key point to remember is that all **breastmilk is good**, and that **milk volume counts more than fat content**. Discussions of how long each side are irrelevant for most mothers.

It is true that the first few minutes of a feed provide slightly more watery milk, known as foremilk, in case the baby just wants a drink. Then the fat content of breastmilk increases quickly, ending up with high fat and low volume hind milk.

While this might be interesting, **most women never need to worry about it. Offer your baby all the milk they need**. When you are getting started, it is common sense to offer milk from both breasts each time.

Current advice is to feed on one side until baby is not making good use of the time, despite all your efforts, and then offer the other side. Start with the fuller side next time. You can get apps to track this, or else gently squeeze each breast and you can feel which one is fuller.

Only mothers with a very fast flow of breastmilk, or copious supply, might be advised to restrict their baby to one side each feed. A skilled lactation consultant may suggest block feeding (just on one side for a period of time) to deal with oversupply and sometimes to reduce reflux symptoms in the baby.

How do I know when my baby is full?

In the early days you don't know, you will have to feel your way. "Normal" feeds can be anything from 5-10 minutes to 45 minutes.

Your baby may stop sucking and fall asleep for several reasons, including being full of milk. However, your baby might also have just got tired, or not been receiving much milk at all because supply is low, or maybe they have a sore head from an instrumental delivery or a tongue-tie, or perhaps they were sleepy with jaundice or from being premature… it can be hard to know.

Ask yourself, how long was your baby feeding for? Did you see active sucks and swallows? Is there another risk factor that might make your baby reluctant to feed well?

If you think your baby might still need more milk, try stimulating them. You might squeeze the breast tissue to cause a let down of milk, talk to or tickle them, take off some of their clothes, change their nappy or wind them, then try again.

If your baby is still not willing to feed on that side, try swapping sides at this point to offer a fuller breast. Still no luck, then see how long your baby stays asleep for. If baby is awake and hungry after 20 mins then perhaps that indicates that you need to aim for a longer feed next time. If baby sleeps for 1.5-3 hours, then this indicates that they were full.

Many young babies look and feel full after their feed, then wake up ten minutes later wanting some more! In most cases this is totally normal, just go ahead and offer them the breast again. If your little one clearly wants to suck but is writhing around and unwilling to relatch, try winding them.

If something seems unusual, or your baby appears to be in pain during or after most feeds, please get some help from a lactation consultant who can help you to work out what is going on.

Do I need to wind/burp my baby?
If your baby shows signs of discomfort, you could try winding them.

Some babies may need to burp or fart to enable them to settle. When parents encourage them to do this, it is called "winding."

Babies that need to be winded often stop what they are doing, go a little red, sometimes stiffen their back or raise their knees, and look agitated or cry. To pre-empt discomfort, many parents routinely wind their baby part-way through a breast or bottle feed as well as afterwards.

It is not entirely clear where wind comes from. Both breast and bottle fed babies may swallow air, some more than others, often when they are crying as well as when they are feeding. They also make gas in their digestive system that might need to come out as a fart.

To wind your baby, simply hold baby upright with a straight back, for example resting on your shoulder, and gently rub or pat your baby's back. You might also like to try "wonky winding" in a similar position but with your baby's right cheek against your right cheek and with them slightly diagonal across your body. If there is wind, then your baby is likely to burp or fart before long. If your baby does not burp and looks comfortable there is no reason to persist beyond a few minutes.

If your baby is sound asleep after a feed, as nature intended, I would be reluctant to wake them up to wind them. You may just have an awake and unhappy baby who might need to feed again to settle down to sleep!

Do I have to keep my baby upright after feeds?

Some people cuddle their baby upright after feeds, especially if their baby is prone to wind or bringing vomiting up quite a bit of milk.

There is not much evidence suggesting that we must hold our babies up after feeds. It feels like common sense, because babies like to be carried this way and they seem to keep their milk down better if they are upright. Opinions differ in how long to do it for. Fifteen minutes seems reasonable. Wearing your baby in a well-fitted sling does the same job.

How long should a breastfeed take?
Normal feeds could be anything from 5-10 minutes to well over 45 minutes!

Of course you want certainty in this crazy, unpredictable time, but I cannot possibly say how long your feeds should last, or how long a baby "should" go in between feeds. Nor should anybody else. Baby books, midwives or health visitors that insist on a certain number of minutes on each breast are talking nonsense. There are several, obvious reasons for this. Think about…

- **How hungry is your baby right now?**
  Sometimes a baby is hungrier than at other times. Does your baby want the equivalent of a quick cuppa, or are they in the mood for a three-course meal?"

- **How much milk do your breasts produce?**
  Overall, and today specifically, and what time of day? Most women find their production is naturally highest in the early hours and through the morning, and lowest in the evening. I am guessing that from an evolutionary perspective, nighttime was generally a quiet and safe time to nurse.

- **How quickly does your milk let down to your baby?**
  Some breasts work like fast watering cans, others gently drip milk.

- **How fast and effectively does your baby drink?**
  This can depend on maturity, size, oral anatomy, sleepiness, birth injuries and more. It changes with time.

- **Have your optimised your positioning and attachment so that your baby can get the milk that is available?**
  Kinked milk ducts and shallow latch are not going to help.

- **Do you give your baby the opportunity to have both sides?**
  This is usually a good idea to maximises volume. If your baby is generally contented and feeding well, please do not worry about foremilk and hindmilk. For more on this see "When should I switch sides?"

- **What is your personal storage capacity?**
  It varies even between breasts in one woman.

- **Is your baby having a growth spurt?**
  At times your baby will ask for more milk than usual to meet their developmental needs and boost your milk supply. One of my favourite feeding gurus writes "Common times for a growth spurt are the first few days at home, at 7-10 days, 2-3 weeks, 4-6 weeks, three months, six months and nine months." So, anytime really.

The upshot is that a normal feed can be anything from a few minutes to about an hour. Sometimes several feeds even run into one big marathon session! Be reassured: how long your baby feeds is largely irrelevant as long as your baby is getting fed and you are able to cope with the demands being made of you.

If you are concerned that your baby is not getting enough milk, see the sections on milk supply and warning signs.

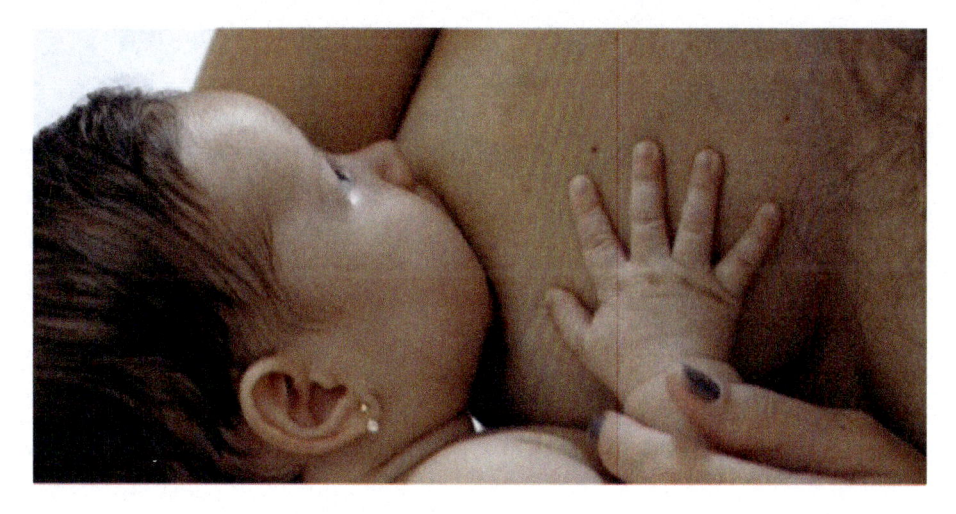

What are the signs that my baby is feeding well?

Many parents worry about how much milk their baby is taking in the early days.

Your baby is likely to be feeding well if:

- Baby has 8-12 feeds in 24 hours.

- Baby feeds for between 10-45 mins at each feed.

- Baby has normal skin colour.

- Baby is generally calm and relaxed while feeding and content after most feeds.

- Baby appears well, alert, and happily awake at several points in the day.

- You can see and hear sucking and swallowing frequently during the feed.

- Baby sleeps between most feeds for 1.5-2 hours, and wakes for feeds naturally.

- Baby is doing an appropriate amount of wee and poo. The number of wees and poos goes up from birth. From day six, babies should do six big wees and about four poos a day.

What should I expect in my newborn baby's nappies?

Your baby's wees and poos give you the best indication of how feeding is going.

Your baby needs at least one poo (black meconium) and one to two big wees (45 ml) in the first 24 hours. If you are not sure whether your baby is weeing, try putting a piece of cotton wool in the nappy (which will stay wet), and feel whether the nappy is puffy or heavy compared with a new one. Many brands put a line on the outside of the nappy that changes colour when the baby wees.

By the third day the poo should be going green. Expect at least three wees and two big poos.

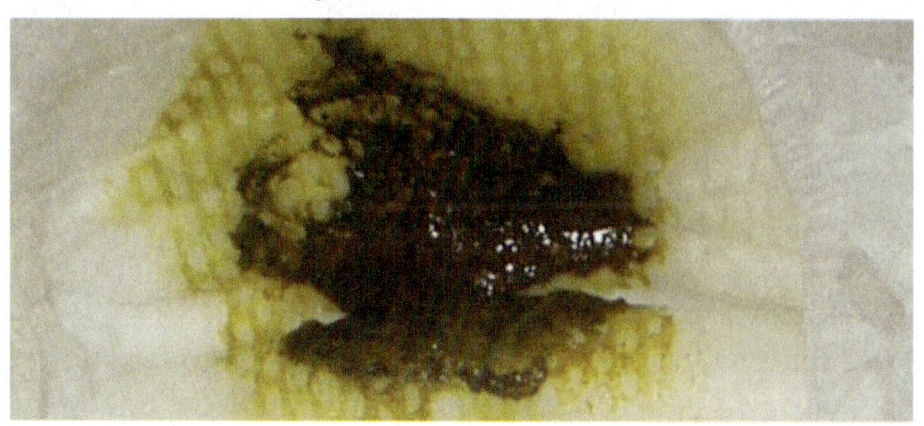

By days four to five expect at the very least two big yellow runny poos and at least five wees. Formula fed baby poo can be slightly browner and pastier but still soft.

Concentrated urine, red dust or crystals in a nappy can be a sign that your baby is not getting sufficient milk, particularly if they happen over 48 hours after birth. Please seek help from your midwife.

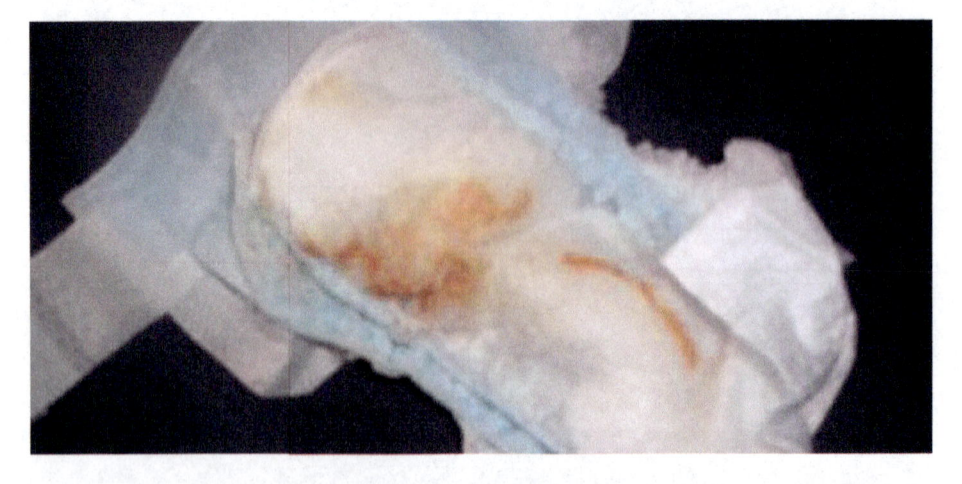

Blood spots coming from a baby girl's vagina in the first week are typically a normal response to her mother's hormones.

Once your baby has six heavy wet nappies a day and about four poos or more, you can relax! We would hope to see this by day six.

What weight gain is normal?

Your baby will be weighed at birth, and then at day five. It is considered normal for a new baby to lose up to 8-10% of their birth weight in this time, as they are receiving small feeds of colostrum. Your baby is expected to be back to birth weight by the end of the second to third week. After this most babies put on 140-200 g or 5-7oz a week.

In the UK you will be given an actual or virtual red book before you are discharged from hospital, this has growth charts on which you can plot your baby's weight gain.

During growth spurts your baby will need to feed more frequently. Allow your baby to do this to boost your supply. You will settle back into your normal pattern after a few days. If you were to give your baby previously expressed milk or formula during growth spurts your breasts do not get the message to make more milk.

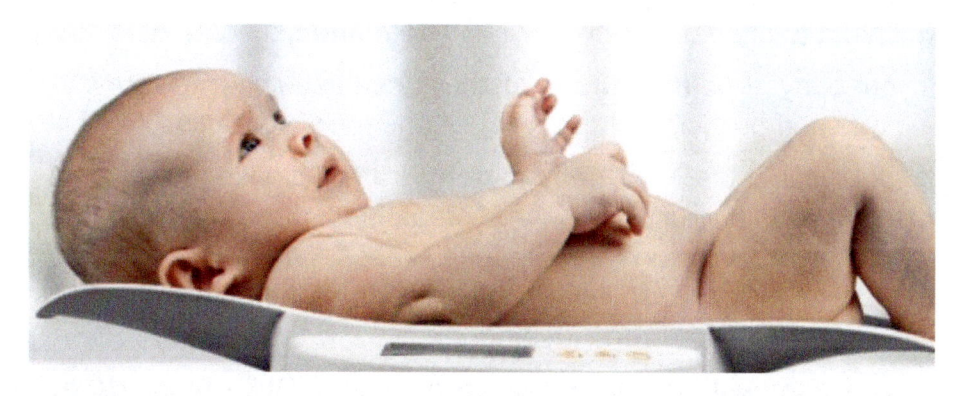

What feeding signs should I worry about in my newborn?

Talk to your midwife as soon as possible if you are worried.

Feeds should generally last 10-45 minutes. During the feed you should see your baby's jaw drop, a slight pause, and then hear them make a "cuh" noise as they swallow milk. If your baby is feeding for **much longer or shorter times** than this, or you are not confident that they are swallowing, you need help.

Your baby should be feeding 8-12 times in 24 hours. These feeds might not be perfectly spaced out, but your baby should not be sleeping for long stretches. A newborn baby might do one sleep of four hours in a 24-hour period, but the rest should be shorter. Ideally your baby will feed, look satisfied, go to sleep and wake up hungry in a reasonable timeframe. **If your baby will not sleep after feeds and still looks hungry, or cries incessantly, this is not normal.**

A very hungry baby can sleep a lot. If your baby does two or three sucks and falls asleep, that was not a proper feed. Wake them up and try again.

If your baby refuses to feed or comes off the breast a lot, this is not a good sign. You know your baby is feeding well when they are doing appropriate numbers of wees and poos. Their wees will be light coloured.

If your bottle-fed baby coughs, chokes, refuses the bottle, dribbles excessively on the bottle, or takes very little milk you will need to change the bottle or flow rate and get some help.

If your baby **appears jaundiced,** they need to be checked by your midwife and possibly in hospital Accident and Emergency (A&E).

If you suspect that your **baby is dehydrated**, please seek urgent medical help. If your baby cannot breastfeed, they need milk from a bottle as soon as possible. In many cases, one or two bottles of expressed milk or formula will help to restore energy levels and allow you to restart breastfeeding.

If your baby will not drink from breast or bottle, please do not delay ringing the hospital or emergency services for advice, even in the middle of the night, or attend hospital A&E.

Urgent signs – seek immediate medical attention and call 999, 111, or go to nearest A&E

If you are seriously worried about your baby for any reason, seek urgent medical care by ringing 999 or 111 or going to A&E at hospital. If you are not sure, they would still much rather see you for reassurance rather than delay.

Signs of an unwell or dehydrated baby include:

- Drowsy, hard to wake, floppy, unresponsive or unconscious **(call 999)**.

- Very pale, blue or grey **(call 999)**.

- Having trouble with their breathing, or breathing irregularly **(call 999)**.

- Ribs and chest pull inward when inhaling.

- Makes soft or-high pitched sounds while inhaling.

- Weak, ineffective coughing.

- Will not suck from breast or bottle.

- No wet nappy for more than 6 hours (after the third day).

- Less than six feeds in 24 hours.

- Concentrated or smelly urine, or orange/red crystals in their nappy after the first day or two.

- Dry mouth and lips.

- Sunken fontanelle (soft spot on the top of baby's head).

- Skin that looks "baggy" and stays pinched-looking if gently squeezed.

- Cold or discoloured hands or feet.

- Sunken eyes.

- Cannot seem to keep any milk down and/or is projectile vomiting (not just occasionally), or has diarrhoea.

- Temperature over 38°C **(go to A&E)**.

When might I need to top up with expressed milk or formula?

If your baby is consistently not feeding well at the breast, and not showing signs of being full, or not gaining weight appropriately, you might be encouraged to offer more milk as a "top up" after a breastfeed.

Top ups can be indicated for various reasons. Some women are not making sufficient milk, and this might be temporary problem or a longer-term issue. Please refer to the section on milk supply.

In other cases, your baby might not be taking milk well due to jaundice, illness, prematurity, a tongue-restriction, feeling tired or sore after a difficult birth, or any number of other issues. Work with your healthcare providers to plan for the hours and days ahead.

A new breastfeeding mother will want to maintain or build her milk supply by expressing milk while her little one cannot access it directly. This can be fed to the baby for their top up, sometimes alongside additional formula as required.

Please refer to the detailed section about giving top ups and how to wean baby off them when appropriate.

How much formula milk does my baby need as a top up?

Your midwife or doctor will guide you to the appropriate amount for your baby.

In their first two or three days of life, all babies have little stomachs that will not immediately stretch to tolerate very large feeds. Towards day four onwards, babies can tolerate more milk (perhaps 45-60 ml). Your midwife or doctor will guide you to the appropriate amount for your baby.

If your baby is breastfeeding and being topped up, we rely on how your baby is behaving to show us how much top up they need. I am reluctant to suggest absolute amounts for any bottle fed baby, because in my experience baby led feeding is more successful than feeding pre-defined amounts. Look at how your baby is behaving. If you finish a bottle feed and your baby is clearly still hungry, showing feeding cues and refusing to settle, you might try offering more milk.

Babies can show feeding cues when they are tired and just want to suck something. You could try settling your baby in another way, for example sucking on your finger or a dummy and see whether they are still asking for more milk after a few minutes.

If your baby routinely spits up a lot of milk after a feed, it is possible that they had a little too much. However, some babies just spit up regardless.

Note that formula milk manufacturers tend to suggest fewer, bigger feeds, even for newborns. Many babies could not comfortably take the recommended amount in one go and would prefer smaller, more frequent feeds.

Should I plan to feed when my baby wants to, or apply a schedule?

Healthy babies that are feeding well do not need a feeding schedule.

There are times when a feeding plan may be necessary. Premature, tiny, unwell or jaundiced babies, for example, might sleep for hours on end if we let them. Clearly, they need to be woken and fed. Parents of twins and triplets often benefit from a schedule, we have a section on this later.

When I teach antenatal classes I am invariably asked when to get a new baby into a feeding and sleeping routine, there is more on this in the section When can I get my baby into a routine? New babies are not capable of learning routines. Trying to apply a routine can result in low milk supply and an unhappy baby, with new parents feeling like failures or looking for solutions to problems that are not even there.

Should I wake my baby to feed?

This depends on your baby. If little one is full term, happy and healthy, and not displaying any warning signs, you can probably be led by your baby. If you feed your baby whenever they are showing feeding cues, and that is roughly 8-12 times a day, then you will get in sync with them and don't have to think about it too much.

There will be other situations where you will absolutely want to keep an eye on the clock and wake your baby if necessary. Tiny, premature, unwell or jaundiced babies might try to sleep for seven or more hours in a row if you let them. Obviously, this is not good for them. For these babies, you might decide to set an alarm for say three hours after the start of your last feed, and keep feeding to a schedule until your baby is big, robust and well enough to let you know when they want to be fed.

Even newborn babies can sleep for a four to five hour stretch once in a 24-hour period. If they do this, wake up hungry, and feed more frequently at other times, that is fine. If your baby wants to do this in the daytime hours, it is possible that you choose to rouse them and hope that they will learn to sleep longer at night instead.

What is cluster feeding?

Babies sometimes feed a lot at a certain time of day, or when they are having a growth spurt. This is their way to fill their tummy and boost milk supply.

Many of us assume that our babies will feed a reasonably large amount at predictable intervals and sleep well in between. This leads us to believe that if our babies do not follow this pattern, there is something wrong with them or us! Nothing could be further from the truth.

Most babies will have times of day when they cluster their feeds. It is commonly in the late afternoon or evening, but can happen at other times if the baby is having a growth spurt or developmental leap, isn't feeling well, needs comfort or any number of other reasons. At these times, your baby just wants to feed, and the instant you try to put them down they start rooting around or screaming to feed some more.

Many mothers feel that their breasts are soft and empty and assume that they need to give their baby formula to satisfy their hunger, but that is rarely helpful in the long term. Breasts are never empty; you cannot drain them because they are always making milk. When a baby's feeds seem to roll into one, they are getting fat-rich milk and stimulating the breasts to produce more milk overall. That is how the supply and demand equation works. Lots of babies who cluster their feeds all through the early evening are then able to have a good sleep afterwards.

Formula and mixed fed babies are not immune to cluster feeds, they too want the comfort of a feed and might also need to show you that they are growing and need more milk in their bottles.

I am not suggesting for a minute that cluster feeding is fun for a breastfeeding mother. Very often, she will be at her wits' end. It can be helpful to remember that clustering is most common in the first 12 weeks (especially the second and third night after birth), and that many babies (though not all) do stretch out the time between feeds as they get bigger.

Mothers suggest several ways to cope with cluster feeding:

- Lower your expectations of what you are going to get done.
- Welcome the chance to snuggle with your baby and heal your body.
- Take any offers of help you can, especially with shopping, cooking, cleaning and other domestic tasks, and looking after older children and pets.
- If nobody offers, ask for help!
- Create a lovely environment to feed in – surrounded by everything you need.
- Consider feeding lying on your side in bed.
- Perhaps try feeding in a well fitted sling.
- Have lovely nutritious drinks and snacks handy. Drinking bottles with a built-in straw can work well.
- Try all those books, films and podcasts that you never generally get round to.
- Use the time to tick off one of the tasks that has been niggling you, like paying a bill or sorting out the photos clogging up your phone.
- Reach out to any friends or family who understand.
- It is OK to vent your feelings, this is hard work.
- If necessary, leave your baby with someone and get out of the room or even the home, to clear your head.

Should I try to stretch out the time between breastfeeds?

"I read that my baby should be feeding every four hours by now, but he is hungry much sooner than that. How can I stretch out the time between feeds?"

There is no need to put your baby into a strict feeding routine in the early days (unless told to by a medical professional). There is no requirement to try to stretch out the time between a breastfed baby's feeds especially if it results in a lower milk supply and an unhappy baby.

Here is the science. When breasts are fuller, milk production slows. When breasts are emptier, they make more milk. A baby that feeds frequently from emptier breasts receives milk with a higher fat content. Mothers often worry about evening cluster feeds when their breasts feel empty, but this is exactly what is designed to happen. Their baby is filling up on milk to have a good sleep later. Giving formula at this point is an obvious response in a desperately tired and worried mother, and yet it is rarely necessary from her baby's point of view.

Women's breasts store milk between feeds. The storage capacity is not directly related to breast size and can vary hugely between women. Some babies need to feed frequently because their mother does not store much, and that is just the way it is. Increasing the gap between feeds for these babies can result in a dramatic decrease in milk supply and a hungry baby.

Most women can trust their babies and their bodies. There is no point in comparing yourself to your friends or measuring your baby against something in a book.

When can I get my baby into a routine?

Routines are not necessary for a baby (but they might be for you!)

Most parents these days assume that they need to get their baby into a routine. It sounds like such a wonderful idea: just apply a set of rules, and your baby will become happy, predictable, and sleep much more at night.

Having an unpredictable new baby is totally exhausting and it is almost impossible to get anything done. This leaves new parents feeling like failures or looking for solutions to problems that are not even there. Of course you want some control back in your life. It is only natural. Every day I hear the frustration of new parents who are so tired that they would do anything to get even three or four hours of uninterrupted sleep.

The bottom line is that new babies are not capable of learning routines.

Lots of baby books and maternity nurses want to sell you this beautiful idea, but it is not true or even particularly kind to your baby. Your little one needs loads of reassurance and cuddles and frequent feeds day and night, until gradually they need less of them, and one day they will learn to space out their feeds a bit, until in the end they might sleep a bit longer and maybe all night sometimes. This all happens by itself, when your baby is ready. You do not need to do anything apart from respond to your little one.

It is true that some babies start to get into some sort of rhythm to their day from about six weeks, and even more so from 12 weeks. Hanging in there until 12 weeks without worrying at all about a routine is a good idea.

Yes, you can do little things to nudge your baby towards recognising day and night. Many parents try a set time to start the day, and the research suggests that this can be beneficial. Some parents try to keep night feeds a little less stimulating, in low light and with less distraction. Some try to get into a bit of a pattern of say, "feed, bath, feed, bed" at a particular time. Many parents might try to wake their baby if they are having a long sleep (four hours plus) in the day, with the hope that they might get more calories into their baby in daylight hours. After your baby is a few weeks, a consistent morning wake time helps many babies and parents. You could try these things, but you don't have to.

There is not much evidence that a bottle of formula before bed makes a baby sleep longer. Although many parents like a partner to offer an evening bottle or expressed milk or formula so that mum can crawl into bed for a much-needed rest. Whatever works for you.

In the meantime, please get help. Please recognise that a new mother is designed to hang out with her baby and that it is everyone else's job to look after them both.

Being dependent is a hard thing for a modern woman to get her head around. Here is the thing. Although it feels like you are not achieving anything, you are doing the most important job in the world, 24/7. You are helping your tiny infant to adjust to the world, showing them love, nourishing them, helping their brain and body to grow, investing in one of the most important relationships in your life. Your own body is healing, slowly returning to normal and adjusting to motherhood. Will you be cross with your baby and the massive demands that they put on you? Probably. Who wouldn't be? And when they learn to smile at about seven weeks and beam at you when they wake for the fourth time that night, will you forgive them? Of course you will.

What is it like breastfeeding from three weeks to three months?

Wow, you might be over the hardest part! Hopefully by three weeks your baby will be over birth weight, you might have got more than two hours' sleep in a row, and you might have contemplated leaving the house for more than a few minutes.

For mothers recovering from caesarean section, I hope that you will be feeling less tender and more mobile by now, but still take it easy as you have had major surgery and very little time to recover. Mothers with perineal or internal trauma also need more time to get better. Even after the most straightforward vaginal birth, a woman's body has a lot of shifting and healing to do. Be kind to yourselves.

Hormonally, you are likely to be all over the place. The first two weeks are hardest, but you are still likely to cry every day about something. Tiredness (and resentment) build up. You might be beginning to realise what a huge transition has occurred in your life, especially if you're now looking after your baby alone for much of the day.

You might like to attend local support groups for mothers, such as those run by the NCT and La Leche Leage or the Association of Breastfeeding Mothers. In many areas you have access to Local Authority or NHS centres that might include specialist feeding help or peer support. There are also national helplines to call day and night.

In terms of breastfeeding, most mother's questions now centre around the following topics, and they are covered in this guide. I hope it sets your mind at rest and/or prompts you to get appropriate help.

How does feeding change from four months to six months?

Babies at four months suddenly seem even more interested in the world, and often much more distractable. They may be wakeful in the day and night, and breast or bottle feed in small amounts before getting super interested in something else.

Your breastfed baby's latch might look terrible! They will often squirm at the breast, pop on and off, wriggle and squirm. Many babies will tug at the nipple, squish the breast with their hands, or grab another object like your jewellery or hair.

Bottle fed babies might appreciate a faster flowing teat at this point.

A lot of parents panic about their baby's milk intake, but there is no need if their baby is producing the right number of wet nappies and is still growing.

By this time, many babies can go a few days between poos and then produce a "poonami." If the enormous poo is soft when it comes, there is nothing to worry about.

Very often we are now adding teething to the mix, with babies trying to chew everything and sometimes biting. Nursing strikes are not uncommon. We have sections on both later in the book.

Concerns of parents of babies from three weeks to three months.

Baby:
- Wind.
- Fussiness.
- Colic and reflux.
- Allergy or intolerance.
- Baby not sleeping/routines/wake windows.
- Slow weight gain.

Parents:
- Exhaustion.
- Still having sore nipples.
- Mastitis.
- Thrush.
- Routines and baby sleep.
- Concerns about diet.
- Milk supply.
- Top ups – adding or removin.
- Pumping schedules.
- Not liking breastfeeding.
- Low mood.
- Postnatal depression and anxiety.
- Feeding in public.
- Introducing bottles.
- Taking breastmilk or formula out with them.

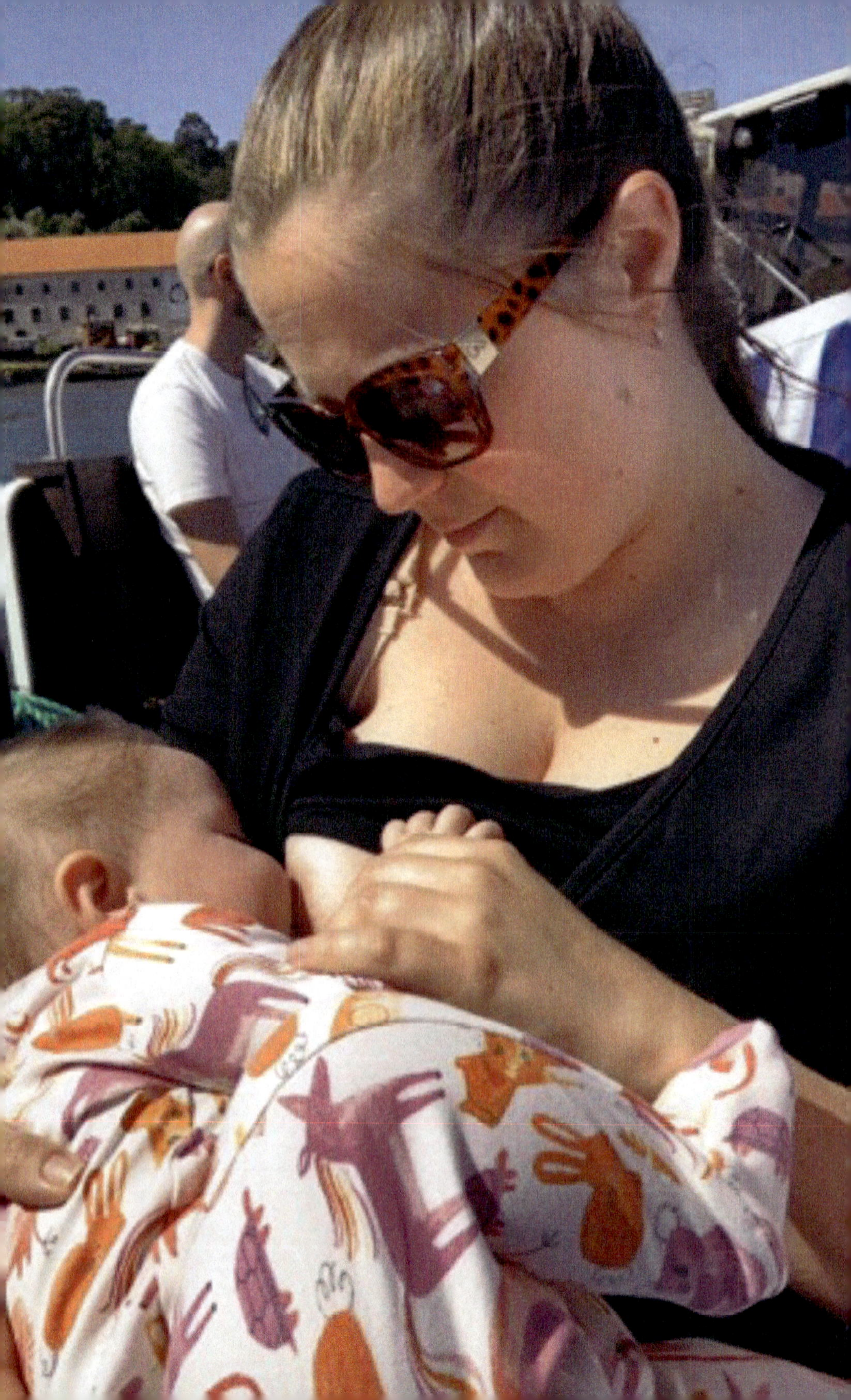

How does feeding change from four months to six months?

Babies at four months suddenly seem even more interested in the world, and often much more distractable. They may be wakeful in the day and night, and breast or bottle feed in small amounts before getting super interested in something else.

Your breastfed baby's latch might look terrible! They will often squirm at the breast, pop on and off, wriggle and squirm. Many babies will tug at the nipple, squish the breast with their hands, or grab another object like your jewellery or hair.

Bottle fed babies might appreciate a faster flowing teat at this point.

A lot of parents panic about their baby's milk intake, but there is no need if their baby is producing the right number of wet nappies and is still growing.

By this time, many babies can go a few days between poos and then produce a "poonami." If the enormous poo is soft when it comes, there is nothing to worry about.

Very often we are now adding teething to the mix, with babies trying to chew everything and sometimes biting. Nursing strikes are not uncommon. We have sections on both later in the book.

Settling your baby

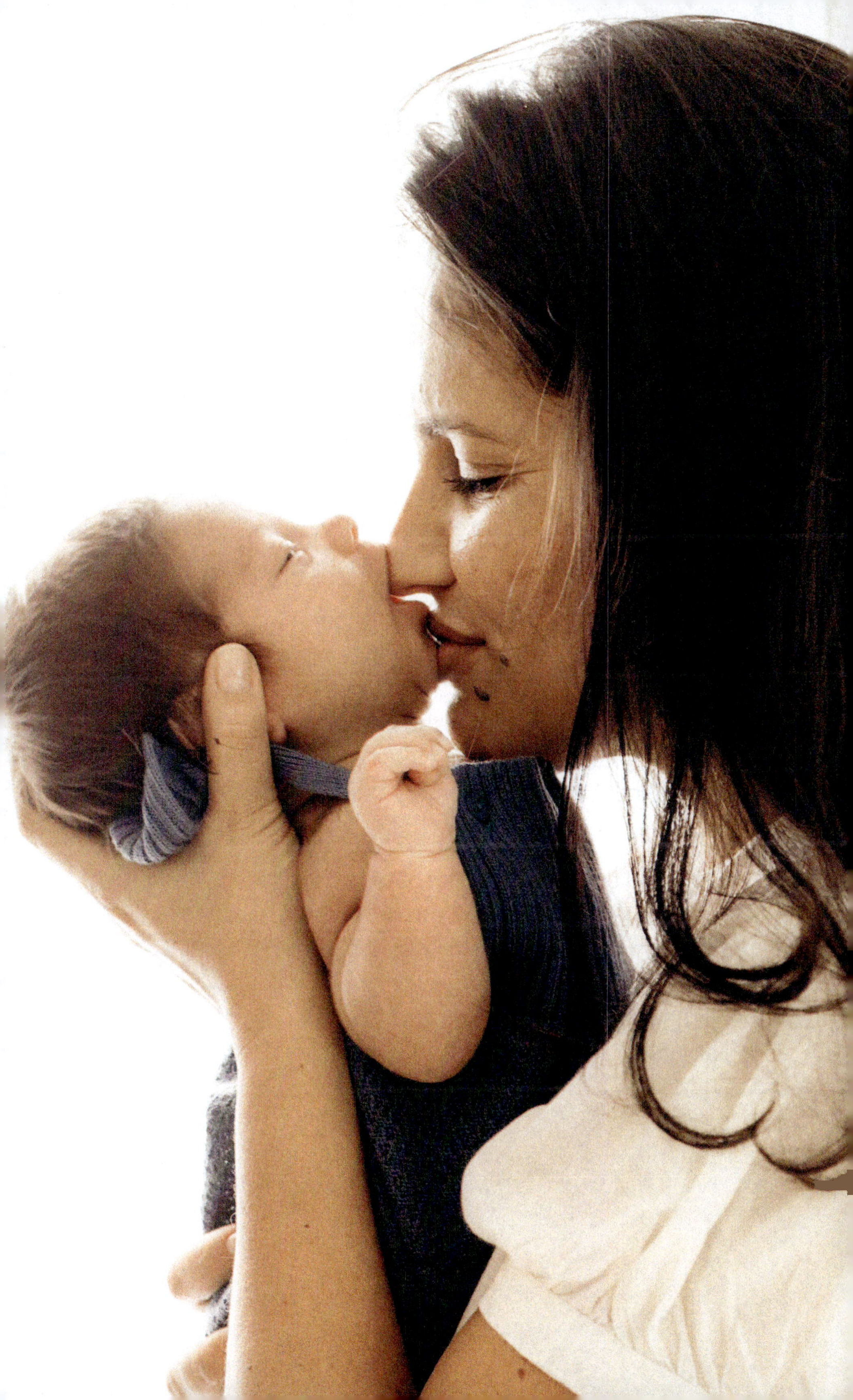

Is it OK to feed my baby to sleep?

Yes! This is completely natural in the early days. Feeding is meant to be cosy and comfortable for parent and baby and is a great way to settle a baby to sleep.

New parents are often told that their baby must learn to "self-settle." Books insist that babies must not associate feeding with going to sleep, in case it builds an unsustainable reliance on the mother. To promote baby's independence parents are even instructed to rouse their softly sleeping baby after a feed before putting them down to sleep, and to structure their babies' lives with "eat, play and then sleep" routines. If their baby falls asleep on the breast or in their arms after a feed, it is considered a failure because they are creating bad habits. But this is total nonsense.

Dr Pamela Douglas, author of "The Discontented Little Baby Book" explains that these feed, play, sleep cycles mess with babies' biology. We are being told to ignore the powerful, biological cues of sleepiness at the end of a feed. She writes, "Disassociating sleep from feeds leaves parents rudderless, because the most dominant and repeated cue that baby is ready to sleep has just fallen out of each day."

Are we making a rod for our own back by letting baby feed to sleep? No, we are keeping in sync with natural processes. Will our baby learn to sleep on their own eventually? Yes, of course, their capacity to self soothe is innate and develops with maturity in the first two years of life. Trying to introduce sleep strategies, especially in the first six months, has no scientifically proven benefit later in life.

Do I have to put my baby down to sleep?
No, new babies often need contact naps.

Babies crave two fundamental things above all else: food, and sensory stimulation. They get food from generous breast and bottle feeding, and sensory stimulation from interaction with their carers and their environment.

It is odd to think that a little baby will want to sleep on their own without the normal noises of daily living and their care givers around. A baby would prefer, of course, to be asleep in someone's arms or against their chest, but if your baby is kept safe and close they will generally be fine.

Can you feed and cuddle your little one to sleep? Yes, it is one of the joys of parenthood. Is it OK to carry your baby around asleep in a sling? Yes, if it is well fitted and designed so that you can see your baby (upright rather than horizontal) your baby will probably love it. What if your baby prefers to nap when you are outside with a pram. Absolutely fine. Are you setting up bad habits? Of course not. You are working with normal primate behaviour rather than against it. It is great to find ways that allow your baby to sleep while you get on with normal daily life. Experiment and see what works for you.

How much sleep should my baby have?
Parents are often amazed to learn that a perfectly normal newborn baby may take anything from nine to 20 hours sleep in a 24-hour period, and that this might vary widely day to day.

Even at six months old, healthy babies can sleep from anything from nine to 17 hours out of 24. That might mean that the baby next door sleeps twice as long as your own baby, and that both babies are completely fine. It also explains why rigid sleep routines and schedules are often completely inappropriate and frustrating for the whole family.

Our baby's sleep, just like our own, depends on two unconscious biological processes: sleep pressure building, and the circadian clock in the brain.

"Sleep pressure" is caused by chemical activity in cells in the body and brain. Sleep pressure builds up while we are awake and dissipates when we sleep. The more tired we are, the longer it takes to go away. Newborn babies get tired quickly and may need very frequent naps and sleeps. Adults can usually cope with longer periods of being awake, but many of us still like a nap in the day now and then.

Our circadian clock consists of 50,000 cells in our brain, and it regulates our bodily functions through the day and night. It is controlled by environmental cues, particularly light falling on the retinas of our eyes. The activities of daily living help to calibrate the clock. We are woken by light, birdsong, the normal noises of a household getting moving. Through the day we respond to cues of sights, sounds and smells, conversations, activities, and these change as we move towards nighttime.

Our newborn babies pick up on all of this too. By two months old they tend to have more of their sleep at night, and by four to six months their circadian clock is more mature. Regardless of whether we do anything. In fact, deliberately shielding our baby from normal noise and light is an odd thing to do. Making them lie in a covered pram or dark room during the day is interfering with the process of setting a normal circadian clock.

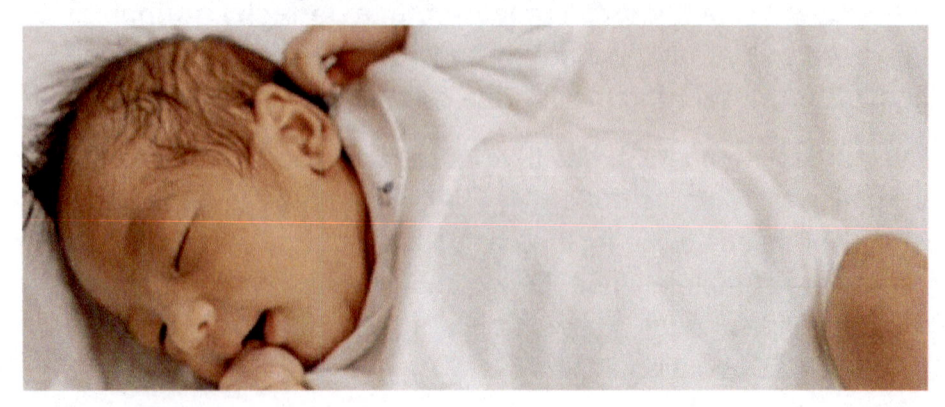

What? I don't have to put my baby down for naps at specific times?
You can if you want, but you do not have to.

We cannot make our babies, or ourselves, go to sleep. We can only remove the obstacles that get in the way. As we have seen, babies need widely different amounts of sleep. If there is nothing preventing sleep, such as unresolved feeding issues, you can get on with your day safe in the knowledge that your baby will take sleep when they need it.

I imagine that you are raising your eyebrows and thinking "*but Gina Ford says…*" There are so many books and programmes out there that suggest babies need structure and routines and that it is our responsibility to get them into patterns, so they don't get "overtired." But think about it. Do you stress that your dog isn't getting enough sleep? If your dog yawns, do you think "quick, I have to get the dog into a darkened room and make it go to sleep"? You trust that if your dog is tired enough it will curl up and go to sleep by itself. It is just the same with your baby. If you have met your baby's needs for a full tummy and plenty of sensory stimulation, your baby will sleep when they need to.

I think about this a lot, because I meet so many mothers who are at their wits' ends trying to get their baby into a sleep pattern. The number of coffee mornings, lunches and playgroups these mothers miss because they were trying to get, or keep, the baby down in their cot. Poor mothers and babies. They would probably both benefit enormously from getting up or out and doing something! Why rock your baby for hours in a covered pram or stay in a darkened room when you could go out for a walk, meet a friend, strap the baby in a sling and make a cake, sit in the park, rather than be a slave to what the book says. Up to 20% of women in the UK experience postnatal depression.

I wonder whether these unnecessary and unrealistic expectations of our babies and ourselves might contribute to this? The way society makes us believe that we must stay at home to be good mothers. Strict routines can result in us striving, overthinking, doubting, comparing, and ultimately not enjoying these precious times with our babies.

But I need a routine!

Ah, now this is a different thing. Many parents are in the habit of running their day by the clock, and the unpredictability of life with a baby can be overwhelming. In your favour, starting and ending your day at the same time, and keeping within your rhythm, will help your baby's circadian clock too. Just please do not expect too much from yourself or your baby.

What about sleep training? Is it harmful?
Luckily for humanity, babies are largely "parent proof."

The sleep training regimes used for toddlers are often applied to new babies, and yet there is no evidence that they work before six months at the earliest. Some people would suggest that they are positively harmful, and there is not much reliable evidence of that either. The book "Cribsheet" by Emily Oster has an interesting review of the scientific data on this.

The difficult thing about sleep training for babies is that parents are told to deliberately delay their responses to what they know their baby is asking for, or not respond at all. Babies have an acute biological need to be close to their parents, and parents are hard wired to respond to their baby. Sleep training, especially "cry it out" methods, means not meeting these needs. This can be very stressful for both babies and parents and could cause miscommunication between you.

Babies need to learn that their needs will be met when they signal them. Once they have learned that, they tend to be calmer overall. Surely our role as parents is to keep things as calm, kind and easy as possible.

When can I introduce a dummy/pacifier?

Dummies can be great if they are kept clean and you do not use them to delay or miss feeds. Current guidance is to introduce a dummy after three weeks if you want to.

Babies love to suck! They have been sucking their thumbs in the womb and would find it enormously soothing to do so again, if only they knew where their thumbs were and how to reach their mouths! Parents often instinctively offer their baby a finger to suck on to calm them down. Some parents prefer to offer a dummy or pacifier in these instances.

If you are certain that your baby is well fed, and you are committed to generous feeding of your little one when you see feeding cues, offering a dummy to your baby can help them to calm down or go to sleep. Many babies in hospital intensive care are given dummies for this exact reason. The balance of evidence suggests that the sucking practice helps them to feel better and feed well later. Some osteopaths recommend dummies for babies born by caesarean section, who need to suck more to keep comfortable after birth.

Some babies never want or need a dummy, and some parents chose not to give one. This is all fine too!

However, new parents are often told not to give their baby a dummy because it can "interfere with breastfeeding." Evidence suggests that it is not the dummy itself that is the problem, it is baby missing out on feeds. If a parent sees feeding cues such as rooting or crying, they might be tempted to give their baby a dummy instead of feeding them, or to delay feeding. These babies get less milk, which is not great for their health and can also reduce the amount of breastmilk available in breastfeeding women. For this reason, many health professionals suggest delaying use of a dummy until breast and bottle feeding is well established (perhaps around three weeks).

Some studies have associated regular use of a dummy with a decreased risk of Sudden Infant Death Syndrome. The mechanism for this is unclear and many doubt that there is a link at all. It might be that babies using dummies are less likely to sleep on their fronts, be slightly more aroused, have an easier job keeping their tongues forward and airways open, or are checked more frequently by their parents. As this is all uncertain, dummies are not recommended to all parents on this basis.

A downside of dummies is the need to keep them clean. They may be vectors for bacteria and fungi and are correlated with an increased risk of infections, especially middle ear infections. If using a dummy, please make sure that it is properly cleaned at regular intervals according to the manufacturer's instructions. Many dental professionals recommend getting rid of the dummies before your baby is two years old to avoid impacts on the positioning of their teeth.

So, use a dummy if you wish to. Just do so with caution, and certainly not to delay feeding your baby.

Why won't my baby sleep in the cot?
Your baby's natural habitat is your body for at least the first month.

So many new parents ring me in desperation in the first few days at home with their newborn. Their baby has been feeding at the breast for what feels like ages, and then the instant they try to put the sleepy baby down in the Moses basket they wake up and look around for more food and cries until they are put back at the breast, and so the cycle continues.

Bottle fed babies often do the same thing, constantly giving feeding cues although they have only just fed. The sleep-deprived parents are at their wits' end and think that everything must be going wrong. **Except it is all perfectly normal.**

Think about it from your baby's point of view. Your little one has been carried around in a warm place, snuggled up, listening to your heartbeat and breathing, for their entire lives. As far as they are concerned, being born is rubbish! They long to be on you, ideally super close and snuggled up skin to skin, where they are the perfect temperature and soothed by your voice and the sound of your body.

Nature has set this up perfectly. Parents and babies are designed to be together! Close contact will boost your bonding hormones and keep your baby calm. Breastfed babies will be able to feed little and often and build mother's milk supply. It will not always be like this, but for the first few weeks you will do well to go with the flow, cuddle and love your baby, and offer the breast or bottle every time they show feeding cues like opening their mouth, trying to suck their fingers and rooting around with their mouth open.

I promise that you cannot spoil your newborn baby. They are incapable of manipulating you. Research shows us that responding to your baby and getting in sync with them builds the foundations for a wonderful healthy relationship and a happy little one. Your family and friends can help by looking after you, feeding you, taking on other responsibilities, and giving you space to get to know your baby.

Of course, it is important to know that your baby is feeding well and not fussing to be picked up because they are very hungry. You will need to keep an eye on how your baby is feeding, and what is coming out in their nappies. If you are worried about how much your baby is drinking, get in touch with your midwife or healthcare provider. If your baby is well fed, but just wants to be with you, then try to relax and enjoy these cosy moments. Most new parents say that the newborn stage was hard but somehow passed in the blink of an eye.

Is it safe to breastfeed my baby in my bed?

Yes, this can be a great way to feed and even co-sleep if you take common sense precautions to minimise risk.

When I talk to new parents about breastfeeding their baby while lying in their bed, they look at me askance and say, "What if I fall asleep?" For years, parents have been told that it is safer to wake themselves up, get out of bed and go to another room, feed their little one in low light conditions, then settle them back in their cot.

However, this is a long way from the reality of mothers who are so tired they can barely raise their heads, desperately trying to stay awake on a sofa or chair, and the drama of trying to put a newborn down in their cot when they know they want to be with their grown up! What really happens? It is estimated that one in five newborns are in bed with their parents any given night, but nobody talks about it!

On the flip side, when a mother settles on her side in bed, with her baby lying next to her feeding happily, both are full of blissful oxytocin. When a mother is tender after birth, it relieves pressure and allows her to relax. For a baby recovering from a challenging birth, there is a lot less holding required to keep them comfortable. The baby feeds until they fall asleep, and nobody needs to go anywhere. The mother can gently move her baby into their crib, or just roll them onto their back and snuggle down. Studies suggest that the whole family often gets more sleep this way. Of course, this is suitable for daytime naps or just relaxation as well as overnight sleeps.

In the early days, you can always ask an adult who is awake to keep an eye on you both as you feed and snooze. This is particularly great when recovering from the birth.

After a caesarean birth, many mothers wait a few days until there are comfortable lying on their side, and they might use a rolled-up towel or small blanket to protect their wound from an unintended baby kick!

What is the evidence? Is it safe to breastfeed a baby in bed?

Just like anything in life, nothing is 100% safe. We can only reduce the risks to acceptable levels. If you think you might fall asleep with your baby while feeding, plan to do that safely. Sharing your bed with your baby can be just as safe as having them sleep in a cot provided it is done consciously and that you follow some simple precautions.

La Leche League has summarised a lot of research in this area. They came up with The Safe Sleep Seven, to help minimise the already small risks to the baby if sharing a bed.

The La Leche League Safe Sleep Seven.

You need to be:

- Breastfeeding.
- Sober (nobody in the bed should have taken drugs, alcohol or medications that could make them drowsy).
- A non-smoker (nobody in the bed should smoke).

Your baby needs to be:

- Healthy and full term, with a birth weight above 2.5 kg.
- Sleeping on their back when they are not breastfeeding.
- Unswaddled and lightly dressed.

You both need to be:

- On a safe surface – a flat mattress (no hollows that baby could roll into) with no gaps and nothing that baby could get wedged or tangled up in (e.g. cords from a window blind, extra-long phone chargers). Never sleep with your baby on a sofa or in an armchair. Don't pull a duvet over you both.

There are additional simple precautions. Partners sleep behind the mother if they are also in bed, the baby is not in the middle of the adults until they are four months old or more. Keep pets and other children out of the bed while your baby is small. Never leave a sleeping baby next to a fire or radiator, or in full sun, or in warm outdoor clothes when indoors.

I have decided to try this! How do I do it?

Check for the Safe Sleep Seven. Lie down on your side in bed with your baby on their side facing you. Most mothers have their head on a pillow and their baby on the mattress. Mothers typically have their underneath arm bent out of the way and their legs curled a bit – we call this the cuddle curl. The baby will be comfy with their head and body facing the nipple that is on the mattress. Most babies simply suck themselves into place! Many mothers use a hand to support or stroke their baby's back. Within a few attempts, most mother and babies can literally do this with their eyes shut. When the baby is full, mothers roll them onto their back to sleep or move them into their cot.

Can my partner feed our baby at night so I can sleep?

It often works for a partner to give a bottle of expressed milk or formula before bed, ideally when the baby is older than four weeks.

Most expectant and new parents are keen to share the process of feeding their baby. Partners often love the chance to bottle feed their little one and support the mother in getting more rest.

There are many positive reasons for sharing the feeding process, but in the early days it is not always the easy option. For a partner to bottle feed a breastfed baby, there will be a process of buying, washing up and sterilising bottles and all the associated kit, and then deciding and preparing what will go in them. For many women, that means expressing milk at times of the day when she could be doing something else, or getting well deserved rest.

Still, this might well be worth it! You might find you can express some spare milk after one or more morning feeds when milk supply is often highest, and your partner can give it later in the evening when supply is generally lower. This can buy you time to do something else, like sleep! It is not a great idea to skip feeds in the early hours of the morning, or to leave breasts full for long periods (over 5-6 hours) at any time of day as this can result in low milk supply.

From a purely biological perspective, it might help your breasts and baby to wait for between 4-6 weeks before pumping and bottle feeding, unless there is a pressing medical reason to do it before that. After the first month or so, the breasts are fully developed for feeding and supply is well established.

By then you have a lot more flexibility in feeding and breasts are less susceptible to permanently reducing supply when you go longer between feeds.

Having a new baby is exhausting! Please consider the myriad other ways a partner can support you and the baby without necessarily doing bottle feeds. These include soothing and cuddling, nappy changing, winding and settling, bathing, massaging and a whole host more.

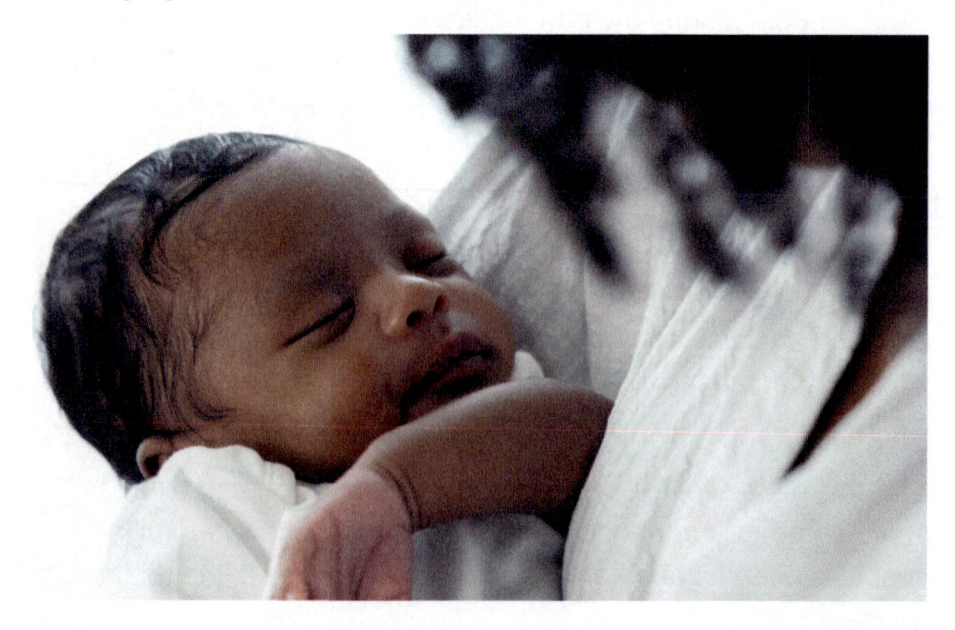

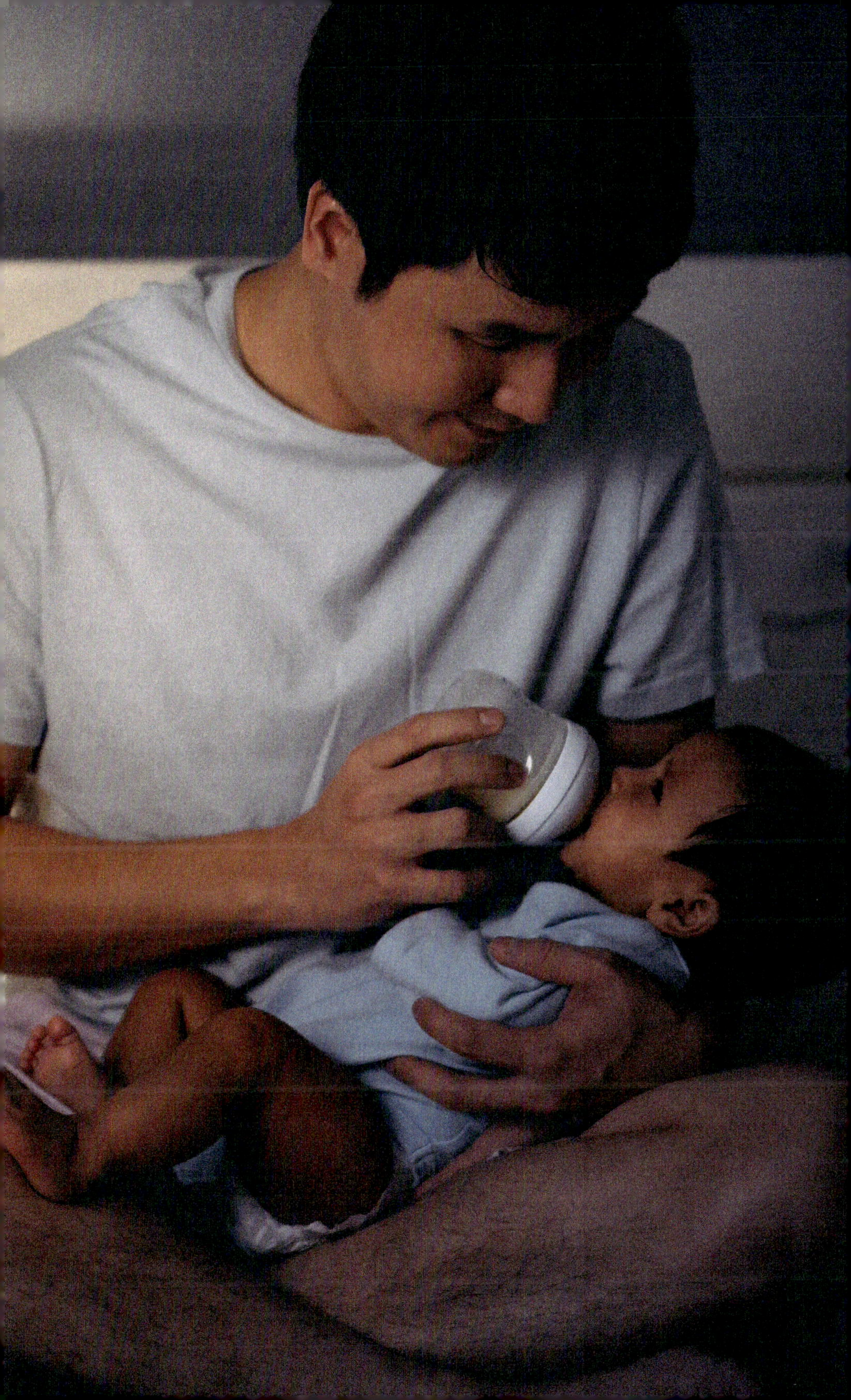

Food, drink and medications

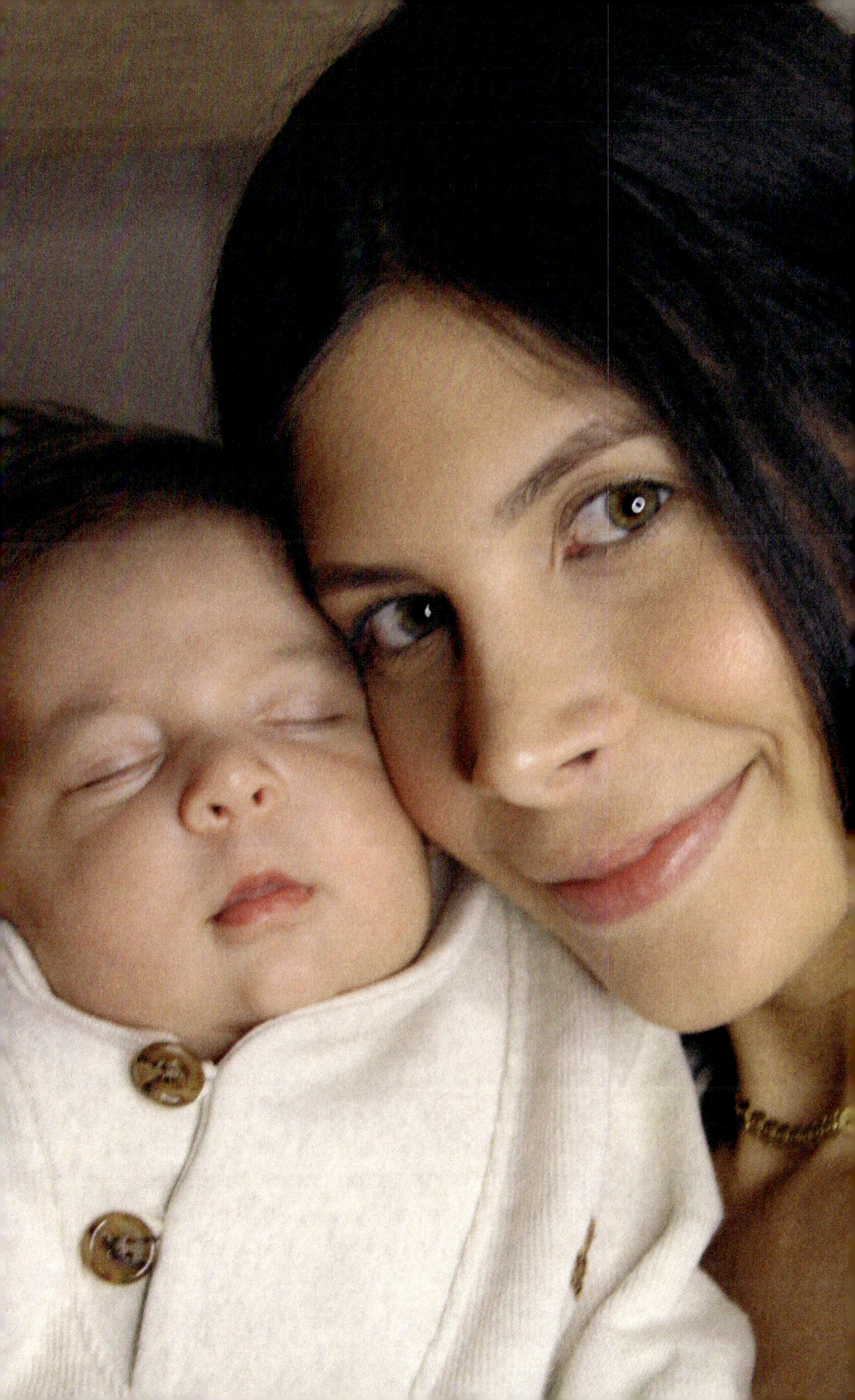

What should I eat and drink after the birth of my baby?

Enjoy a nice balanced diet and drink to thirst. The quality and quantity of your breastmilk is usually fine if you have sufficient calories and hydration. You do not have to maintain a perfect diet.

Whatever kind of birth you have had, it is important to eat and drink well. You will need to drink enough to avoid constipation and allow you to wee sufficiently for the midwife to tick that off her checklist. There is no need to drink to excess when breastfeeding, and in fact this is counterproductive: please just drink to thirst.

I have seen many hospital meals come and go in my time, and they are not particularly appetising. One of the best things that partners, family and friends can do is to bring you yummy food and drink.

The great news is that now your baby is born, you can eat just about anything. There is no real evidence that any food needs to be avoided when breastfeeding unless you or your baby is allergic to it. Bring on the sushi, unpasteurised cheese, rare steak, anything you fancy. The only thing to avoid is tuna/swordfish/king mackerel every day, due to their mercury content.

You will read plenty of articles online that suggest that certain foods make your baby "gassy." While this is a widely held belief, it is not backed up by the science.

There are certain foods, known as galactagogues, that might increase breastmilk production. You will find out more about them in the section "How do I boost my milk supply?" Just because they are "natural" does not mean that they are all safe in significant quantities, or the same quality across the board, so do use them with caution. For most women, a well-balanced diet (including properly supplemented vegetarian or vegan diets) will be just fine.

Can I have alcohol if I am breastfeeding?

The key thing is not to be drunk when looking after a baby, rather than worrying about a tiny bit of alcohol in your breastmilk. NHS guidelines suggest that breastfeeding mothers can drink small amounts of alcohol occasionally, although they should not then bedshare or fall asleep on a sofa with their baby. Heavy or binge drinking is not recommended.

After an alcoholic drink, a woman's blood contains some alcohol, and it passes freely into the breastmilk. However, as the woman's blood alcohol level falls the alcohol also leaves the breastmilk. There is no need to "pump and dump" breastmilk after a drink.

Some women aim to have their drink at the end of or just after a feed, and by the time they feed again in two to three hours the amount of alcohol in the milk will be negligible.

If the mother is drunk and her breasts hurt as they are so full, she might choose to express the milk and throw it away. This has dual benefits of keeping her comfortable and keeping up milk supply. The key thing is that a drunk mother is not capable of looking after a baby properly, a sober adult needs to be doing that until the mother is feeling better!

The NHS stresses that binge drinking, where you have more than six units of alcohol in a session, may make you less aware of your baby's needs. It not great for you or your body, and you might like to chat to your GP or health visitor if you do this on a regular basis.

How much caffeine can I have when breastfeeding?

You are advised to limit your coffee intake to one or two cups a day, the same as during pregnancy.

Many new mothers are keen to have caffeinated drinks like tea and coffee to keep them going. A small amount of caffeine goes into breastmilk, peaking one to two hours after consumption. The good news is that babies receive only about 1.5% of the maternal dose of caffeine, so in most cases moderate consumption is unlikely to affect your baby if they are full term and healthy. Babies' ability to tolerate caffeine increases with age.

Premature babies, those under six months and those with other health issues, might be less able to tolerate caffeine through breastmilk. They might be extra irritable, fussy, or have trouble sleeping.

Most health agencies suggest that a mother may safely drink two-three cups of coffee daily (200-300 mg caffeine). Obviously, teas and coffees vary widely. On average, coffee contains about three times more caffeine than black tea.

Please remember that caffeine comes from many sources including coffee and tea, but is also found in considerable amounts in matcha green tea, some carbonated sodas, flavoured waters, energy and sports drinks. Some medicines also contain caffeine, such as some cold and flu treatments, some pain relief and some weight loss supplements. Guarana, Yerba Mate and Kola nuts/Cola are sources of caffeine. Even 50 g of plain chocolate contains about 50 mg of caffeine. How annoying!

Do I need to take vitamin supplements when breastfeeding?

No, unless you have a very restricted or vegan diet, in which case appropriate vitamin substitutes (such as B12) are recommended.

Are there any foods that I should avoid while I'm breastfeeding?

No, you can eat whatever and whenever you like unless your baby has an obvious allergic reaction to a particular food (rare).

Despite what you read on social media, there are no foods that need to be banned from your diet when breastfeeding. It is thought that everyone should avoid eating significant quantities of certain types of fish like swordfish, fresh tuna and king mackerel that are high in mercury.

As already noted, you can have some alcohol and caffeine in moderation, but do not be drunk when looking after your baby.

It is uncommon for a baby to be allergic to something you are eating. However, it is possible for allergens to pass into breastmilk so if your baby is showing signs of an allergic reaction seek medical advice. You might need to try elimination diets for the most likely culprits (for example dairy, soy, wheat, peanuts, egg). In my experience, many mothers restrict their diet unnecessarily. There is more on allergies in the section Is my baby allergic to cow's milk?

It is not recommended to restrict your diet just in case a future allergy might develop. For example, we have no evidence that you need to refrain from eating tree nuts or peanuts unless you are allergic to them yourself.

Can I diet to lose weight while breastfeeding?

Yes. It is fine for breastfeeding mothers to lose weight safely if they follow some general guidelines. Breastfeeding already burns 200-500 calories a day.

Clinicians suggest waiting for about two months after birth before you start trying to lose weight. This gives your body time to settle and heal and maximises breast development and milk production.

If you are dieting, remember:

- You will still need about 1800 calories a day to produce sufficient breastmilk for your little one.

- You will lose most weight by breastfeeding frequently, on demand, and feeding longer than six months.

- Decrease calories gradually, and keep weight loss at less than 1.5 lbs or 0.68 kilos a week. A sudden drop in calories can reduce milk supply if a mother's body goes into "starvation mode."

- Smaller frequent meals are better for you and your milk supply than restricting yourself to two meals.

- Please avoid fad diets (like liquid or low carb diets), or weight loss medications while breastfeeding.

- Moderate exercise is a great idea to help you to build muscle mass and lose weight.

If a mother loses too much weight quickly it is typically her body that suffers, rather than the quality of the breastmilk. If you want to lose weight it is ideal to do it gradually so that you have the energy to look after yourself and your little one properly.

I have a cough/cold/stomach upset… shall I breastfeed my baby?

Mothers that are unwell are generally able to breastfeed their baby, assuming they are fully conscious and aware of what is going on.

Unwell mothers will always benefit from support from other caring adults. There are, however, a small number of conditions and specific medications that would make breastfeeding unwise or dangerous. Please speak to a healthcare provider before taking any medications.

Coughs, colds, flu, Covid 19, tummy upsets and other illnesses you would manage at home are horrible for a mother. However, she will be making antibodies in her body to fight off her illness, and by continuing to breastfeed she will pass these on to protect her baby. This works the other way round too, when her baby is sick the mother's body produces the relevant antibodies to protect her little one.

Good hygiene is essential in these situations, for example washing hands thoroughly before handling the baby or baby kit, and not coughing or sneezing on the little one. If possible, another adult should look after the baby in another room while the mother is sick, just bringing her baby in for breastfeeds. If it is possible for a mother to express her milk into a sterile container, and for another well adult to feed the baby expressed milk from a bottle, this could be a very sensible precautionary approach.

Please check with your pharmacist before taking cold remedies or other medications. They can contain, for example, a lot of caffeine that may keep you and your baby awake, or decongestants that can accidentally reduce milk supply.

If the mother has an active cold sore (herpes simplex) this can be extremely dangerous for her newborn baby, and it is vital that she does not kiss her baby and practices very good hand hygiene. She must not share anything that touches her cold sore including towels and cutlery.

There is a section on medical conditions that impact on breastfeeding later in this guide.

What pain relief is suitable while breastfeeding?

After a caesarean or instrumental birth you will be offered some heavy-duty pain relief. This is typically dihydrocodeine. There are no prizes for refusing it, mothers recover better when their pain is well controlled.

The Royal College of Obstetricians and Gynaecologists suggests that any medications are taken for the minimum possible time. Any prescription drugs have a small risk of side effects for you and your baby if you are breastfeeding. Look out for signs of excessive sleepiness in your newborn. Perhaps the most common side effect of the hospital drugs is constipation in the mother – which is something that you would rather avoid. Some hospitals hand out laxatives or stool softeners for this.

Most hospitals encourage you to use ibuprofen and paracetamol to manage your pain once you are at home. Ibuprofen is best taken with or just after food. Anything with codeine or aspirin is not suitable for breastfeeding mums, so always take advice before self-medicating or if you have any prescription drugs for other conditions.

Can I take antibiotics when breastfeeding?

Yes, most antibiotics are compatible with breastfeeding, but always alert the person prescribing them that you are breastfeeding.

Some people find that their baby might have a slightly upset tummy from the tiny amount of antibiotics that pass into breastmilk. If this happens it should be mild and pass quickly.

Can I take the contraceptive pill while breastfeeding?

Contraceptive devices that contain oestrogen have been linked to reduced breastmilk supply, even when started after six weeks post birth. Progesterone-only contraceptives are a preferred option.

Please talk to your healthcare provider about the best form of contraception in your case. The progesterone-only "mini-pill" might be a sensible place to start. Most women do not experience reduced milk supply when they start taking the mini-pill at least six weeks after the birth and at normal doses. An Intrauterine Device (IUD) like a coil that releases progesterone into the uterus results in less of the hormone in the mother's blood than the mini pill. However, IUDs, patches and injections are a longer lasting commitment.

Where can I find online advice on medications while breastfeeding?

It is always a good idea to check specific medications with your healthcare provider or pharmacist. They can help you to weigh up the risks and benefits of whatever medication you need, and perhaps consider possible alternatives. However, there are also some useful online resources for additional research.

The Breastfeeding Network provides some excellent factsheets, you can access them online and through their Facebook Group.

You can check specific drug compatibility with breastfeeding at **www.e-lactation.com/en**.

The American site **www.kellymom.com** is evidence based.

You will find generic information on the UK NHS website, and those from the main breastfeeding organisations such as La Leche League.

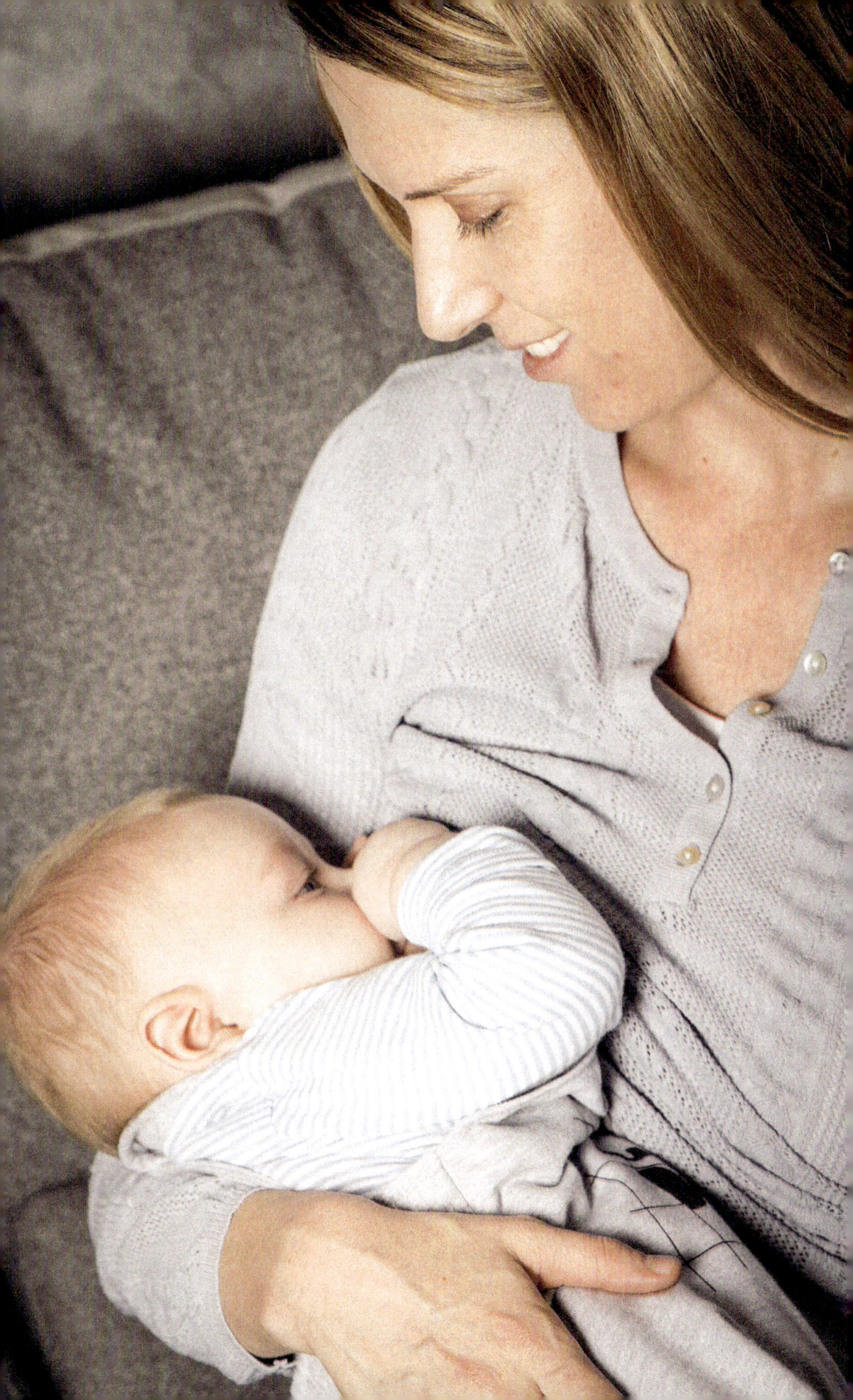

Breast and nipple challenges

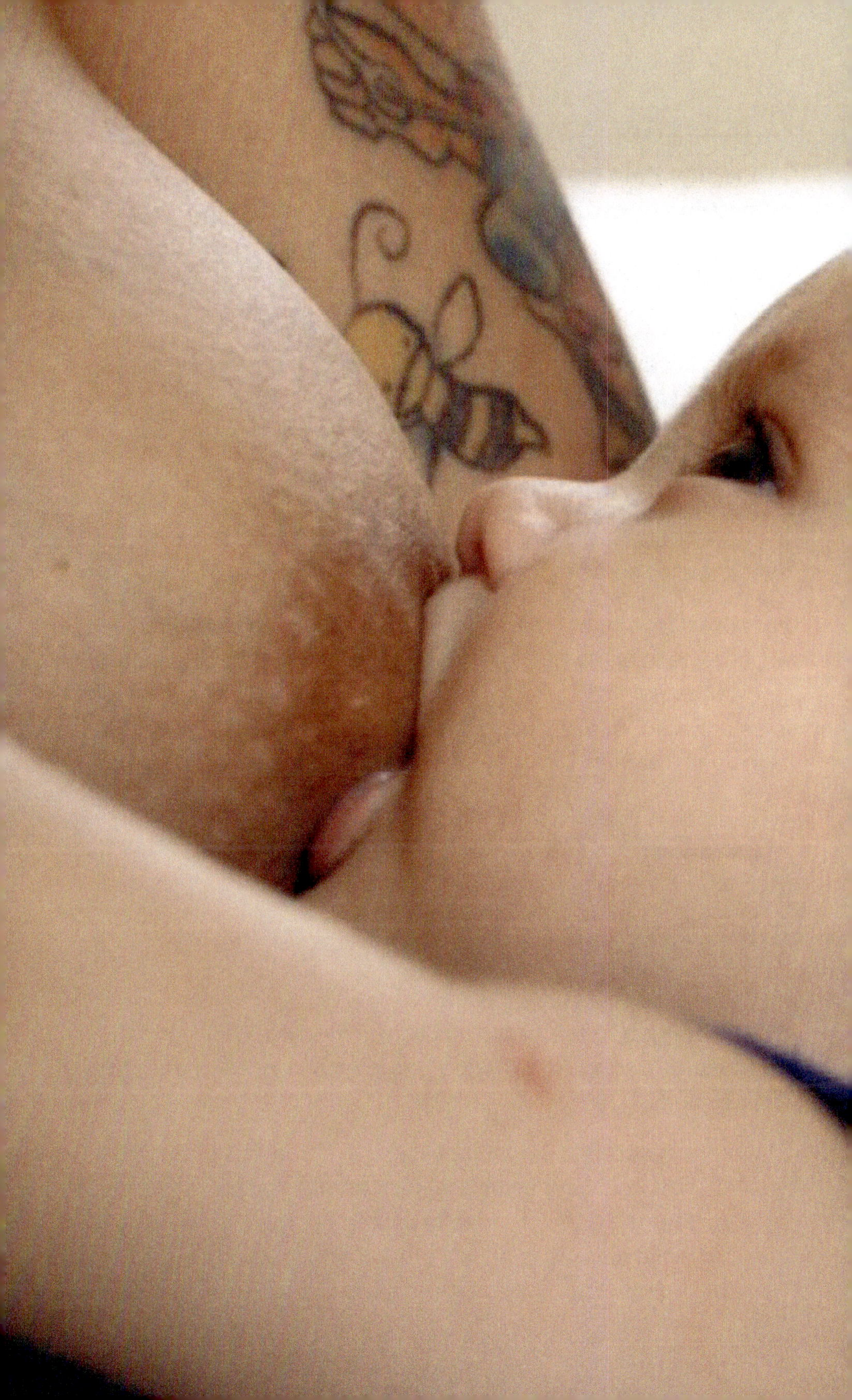

Is breastfeeding meant to hurt?

Not really! Pain is generally a sign in the body that something is wrong and a prompt to do something about it.

"My sister told me that breastfeeding was hideous for a fortnight."

"I read that a correct latch should be painless."

"Everyone says that breastfeeding hurts for the first few weeks and then your nipples settle down."

"The midwife explained that sore nipples are normal because they have never had this much attention before!"

It seems that there is conflicting information on pretty much everything to do with parenthood! Who is right?

I am not denying that many women get sore nipples. However, experience tells me that in most cases we can work together to reduce the pain or make it go away entirely. Once breastfeeding is established, it is designed to be nice. The hormone oxytocin helps both mother and baby to feed calm and relaxed.

Note: the baby on the previous page has a shallow latch and their cheeks do not touch the breast. it is possible that the latch hurts their mum, but some older babies do this and everyone is fine!

Why do many women find breastfeeding painful?

Most often a bad latch, for a host of reasons. Occasionally because of a tongue-tie in the baby, or a condition affecting the nipples.

There are several factors at play here. It is true that most of us are not used to having someone suck on our nipples for up to 12 hours a day, and some sensitivity might be expected. It is also true that while sucking is an instinct for most babies, it can take a baby and mum some time to work out how to get the best fit together.

In my experience, nine times out of ten, a woman gets sore nipples from less than perfect positioning and attachment.

Often the baby is tugging the nipple away from where it naturally hangs, or the baby is not deeply attached so they squash the nipple against the roof of their mouth. Either way, the nipple tends to come out of the baby's mouth looking pinched and can even get cracked and bloody. Not surprisingly, a mother can start to dread the next feed. Sometimes this kind of squishing results in vasospasm, which causes the nipple to go white during and after feeds and gives shooting pains in the breasts even between feeds. Mothers are often misdiagnosed with nipple thrush when incorrect attachment is the issue.

The good news is that a lactation consultant can often help to see what is going on and stop the pain. I have worked with so many women who delightedly exclaim "oh wow, it isn't hurting, this is so lovely" when we find a position that works for them, even when they had been about to give up breastfeeding five minutes beforehand. Please refer to the section on achieving and recognising a good latch.

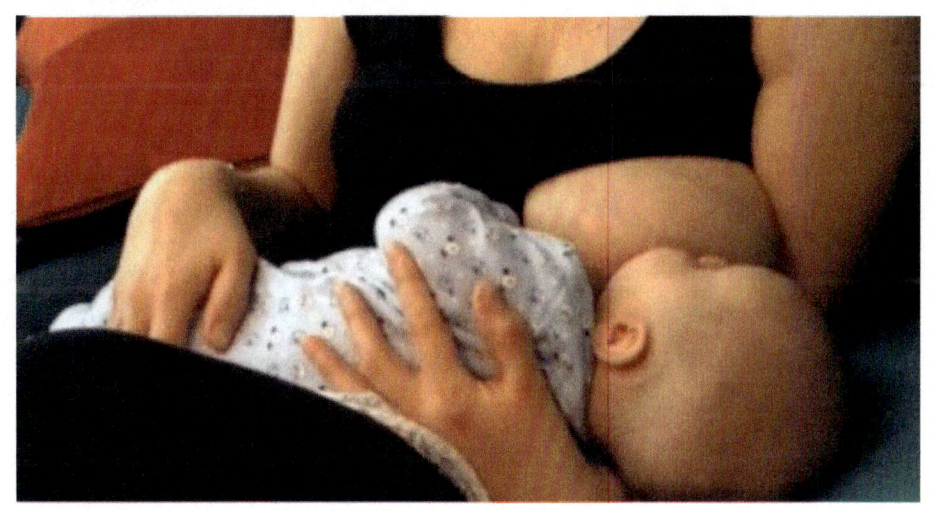

On occasions, the pain is not caused by a position and attachment issue. There are wide variations in babies' mouths and tongue function that can impact on feeding. If you are unsure, please get referred for specialist assessment.

It is estimated that anything up to 10% of babies are born with a **tongue-tie** when the piece of skin (frenulum) from the base of their tongue to the floor of their mouth may restrict the movement of their tongue. These babies may have trouble staying on the breast, and often have "crunchy" shallow sucks that hurt the nipple. While working on positions and attachment can often help, in severe cases it is unlikely to resolve the pain entirely. If you think that tongue-tie might be your baby's issue, please ask your midwife or GP for a referral to a specialist who can fully assess the tongue function, and if necessary perform a simple surgical procedure to release the frenulum. Please see the detailed section tongue-tie.

Some temporary issues also come into play, such as **birth injuries and medications**. Babies that have spent hours being impacted by labour contractions, especially if they were not in a great position, might well have a sore neck. Forceps and ventouse births can result in visible bruises, and many babies are then reluctant to open their mouths wide and jump onto the breast. Little ones may be affected by the drugs that were necessary for mum during labour or after delivery. Many of these issues resolve by themselves in time. Some mothers find that an osteopath trained in paediatric care can help their little one.

I have seen a couple of cases where a tiny baby simply could not fit their mother's larger than average nipple into their mouth. Mothers in this situation are advised to express the breastmilk regularly (approximately eight times a day) and bottle feed the baby until they grow.

Some women suffer from pre-existing breast conditions that can make feeding sore. If you are in any doubt, please see your GP. I have seen a sizeable number of women with **Raynaud's Phenomenon** (a circulation condition where the extremities especially fingers go white in the cold), who find breastfeeding very painful. If you know or suspect that you have Raynaud's you might find that keeping the breasts as warm as possible may help. There are medications available in extreme cases, it would be worth discussing these with your GP.

There are certain nipple shapes and structures that, while normal, can make it more challenging to breastfeed pain free. **Inverted nipples that retract rather than protrude when baby sucks on them, can sometimes cause a mother some pain**. Many nipples look inverted but can be coaxed out, and that is a good sign. However, some are tethered inside and may hurt during breastfeeding no matter what the mother does. I suggest getting help from a trained breastfeeding specialist to work on optimising latch on flat or inverted nipples. Some babies struggle to latch or begin sucking, and there are hints and tips on how to achieve a good latch. Sometimes nipple shields are a good temporary solution.

Breast infections and dermatitis can cause a mother discomfort, and if you are at all worried about what is happening with your breasts or nipples, please seek diagnosis from your medical professionals. **Thrush** of the nipples can result in itchy, scratchy, "broken glass" or "needle" type pain in the nipple and shooting pain in the breast tissue. However, thrush is often misdiagnosed when incorrect latch is causing the problem, and some health professionals doubt that nipple thrush is even a thing. If breastfeeding has always hurt, it is very unlikely that pain is thrush. See the section on diagnosing and treating nipple thrush.

So, if my nipples hurt, I should get some support?
Yes! Absolutely.

Please do not suffer unnecessarily. Most GPs, midwives and health visitors are not extensively trained in breastfeeding. Having a quick glance at you and saying "it looks alright to me" is different from a proper assessment of a breastfeed. If you are in discomfort, please seek out some trained support.

There is so much that trained breastfeeding consultants can do to help you reduce or take away the pain, or at least work with you to understand what is going on and signpost a way forward.

How can I treat my sore nipples?

First, focus on positioning and attachment, and resolving any underlying problems.

Your baby needs a big mouthful of breast tissue and not to be pulling the nipple away from where it belongs. Read this book and consider reaching out for specialist breastfeeding support.

Once you have cracked any latch issues, we can focus on healing. As breastmilk is naturally antiseptic, some mothers squeeze out a bit of breastmilk and rub it onto the nipple area after a feed. Others purchase silver cups that sit inside their bra to protect the nipple and allow it to heal.

Nipple creams, including lanolin ointment, help to prevent scabs forming over cracks. You only need to apply a small amount after a feed, and most brands do not need to be washed off. However, some mums are sensitive to products, so if in doubt stop using them. Wash your breasts as normal when you bathe. Change your breast pads or bra if they become wet with milk.

Many women find hydrogel pads from the pharmacist soothing and healing for sore nipples.

Occasionally, breastfeeding will feel too painful, and a mother might resort to pumping her milk at feed times for a couple of days to let the nipples heal before resuming direct breastfeeding.

If breastfeeding is still painful, please get some skilled support to help find out why. If you are concerned about a nipple infection or dermatitis, please see your GP or pharmacist.

When should I use a nipple shield?

Some babies will not latch despite all your best efforts. Nipple shields could provide an intermediate or permanent solution once the milk has come in (day 2-5).

Shields give a teat shape for your baby to take hold of, so can be a helpful tool for mothers with naturally flat or inverted nipples. You will need to have sufficient supply and flow rate for milk to pass freely into the shield.

Which one is best?

There are many brands of shields, and they are not created equal. In my experience, Medela shields in medium size suit most mothers, although a few will require a larger or smaller size. Many mothers rate the MAM shields, but make sure that you fit them correctly as they have a right and wrong way round in relation to baby's mouth.

How do I use a nipple shield?

Shields need to be thoroughly washed and sterilised before first use, and then washed up well in hot soapy water between uses. When putting shields on it helps to use both hands to partially turn them inside out and stretch out the bottom of the shield. Put the shield onto the centre of the nipple and allow it to draw some nipple up inside as you let go. This makes it easier for your baby to get milk, and prevents the shield from being easily knocked off.

When the shield is on, you would expect it to fill up with milk and to hear your baby swallowing milk while feeding.

If you need to use shields, please do not be put off by well-meaning health professionals who tell you that it might confuse your baby or decrease your milk supply. While these things can be true in some cases, shields can often be super helpful and allow you to breastfeed your baby. If it is a choice of breastfeeding with shields, or not breastfeeding, then use the shields!

Keep an eye on your baby's nappy output and weight gain. If you notice a reduction in milk supply, you might need some additional breast stimulation, for example from expressing. This is more likely if your baby is finding it hard to get milk, for example if they are tongue-tied.

Many babies can be weaned off the shields after a few days or weeks when they are stronger and you both have the hang of breastfeeding. Some mothers do this by starting the feed with the shield, then taking their baby off and removing the shield before proceeding. Other babies continue to prefer shields for weeks or months.

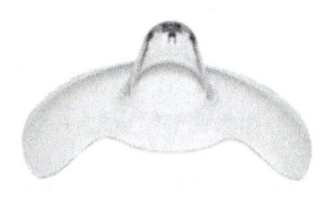

What shall I do about blocked ducts?

Breastfeed or express, paying attention to good milk removal. Consider ibuprofen and cool compresses after feeds.

If you have a blocked duct, you might find a lump or wedge-shaped engorged area in your breast, quite commonly in the part nearest to your arm. The area may feel tender, swollen or hot, and look a little red on light skin tones. You will not be feeling dreadful, as you would with mastitis.

Blocked ducts can be caused by many things that stop the breast from draining properly.

Causes of blocked ducts:
- Suboptimal positioning and attachment.
- Oversupply of milk.
- Tongue-tied baby unable to feed well.
- Infrequent or skipped feeds.
- Pressure on a duct e.g. a mother sleeping on her stomach, or a bag strap or bra that cuts across milk ducts.
- A milk blister (bleb) across the opening of the duct.
- Changes in breast microbiota.

To treat a blocked duct, breastfeed and/or express as normal and aim for good milk removal. Do not go crazy with this, you cannot drain a breast and over expressing will make it worse!

In contrast to former guidance, we now know that ibuprofen and cool help to reduce inflammation of breasts. You are not encouraged to use heat or firm massage, as this makes it worse. Breast tissue is delicate and should be treated with care. Most sensations of blockage will go away on their own.

A cool compress on the blockage might help to relieve the pain. There is no need for antibiotics.

Some women try a few minutes of very gentle breast gymnastics, holding the breast and moving it up and down, then side to side. This should be followed by lymphatic drainage. The woman lies on her back with the arm close to the affected breast resting over her head. She uses her other hand to stroke the breast towards the armpit, with the pressure she would use to stroke a cat. This helps to disperse excess fluid. It is a great idea to feed or pump after doing this.

If the lump in the breast goes down during feeds but fills up again afterwards, this is not usually something to worry about. However, leaving persistent lumps untreated can cause more inflammation in the breast, leading to mastitis (see below) or an abscess. If you are worried about any new lump in your breasts that does not go away, please contact your healthcare professionals.

Some women consider lecithin dietary supplements if they routinely suffer with blocked ducts. The usual recommended dose is 3600-4800 mg lecithin per day, or one capsule (1200 mg) 3-4 times a day. A mother can reduce this dose by one capsule after two weeks if she is not getting blocked. After another two weeks of no blockage, she can reduce by another capsule. This information is taken from the website www.kellymom.com, which provides evidence-based information for lactating mothers.

How can I deal with engorgement?

Feed or express according to your baby's hunger, use cold compresses after feeds to reduce pain and inflammation, ibuprofen and paracetamol can help. Do not express loads of extra milk as this can make the problem worse.

I was lucky to have relatively easy births, but nothing at all prepared me for my boobs on day four. The pain! My breasts were suddenly enormous, super-hot, heavy, stretch marked, they literally felt like they were going to burst. I ran towels under the cold tap and draped them across the melon sized lumps of molten lava on my chest, crying to myself and demanding that my partner sterilise the expressing machine. Eventually, in desperation, I expressed about 200 ml from each side (don't do this!) which felt gorgeous for a while until the whole supply and demand thing kicked in again and my body thought I had triplets. If this sounds familiar, you are not alone. Primary engorgement is common when the milk comes in between day two and day five and can be painful for a few days.

Advice around engorgement has changed recently, and the suggestion is **not** to express more than necessary. Women are encouraged to use cool compresses or smooth cabbage leaves from the fridge to relieve inflammation, alongside ibuprofen and paracetamol if they can take it.

Mothers should **not** massage or vigorously comb their breasts or treat them in any way roughly. This is delicate glandular tissue, and rough handling makes swelling worse. Some women lie on their back with their arm up and use the other hand to stroke gently back across the breast towards their armpit. This can help drain some of the excess fluid in the breast (which is not milk). Mums should apply no more pressure than they would stroking a cat.

Obviously, alongside this we must feed the baby – waking the baby to do so if necessary. Many mums find that warm compresses or a shower help if they are about to feed or express, presumably by increasing oxytocin and helping the milk to flow. If the breasts are too tight for a baby to latch on, some expressing or gentle nipple massage just before a feed can help. There is a technique called reverse pressure softening advocated by many feeding specialists, where the woman uses inward pressure from her fingertips all around her areola before attempting to feed her baby.

The good news is that frequent feeding and getting into sync with your baby should help everything to calm down within a couple of days. In the unlikely event that you become feverish and/or develop angry red patches on your breast (in light skin tones), please consult your healthcare provider in case you are developing mastitis.

These early days can feel horrible physically and emotionally. It helps to have loving support around you.

What can I do about engorgement when reducing or stopping breastfeeding?

Be prepared to express a little if the engorgement gets unbearable and you are at risk of mastitis. Ibuprofen and cool compresses can help.

If you or your baby start skipping breastfeeds or extending the time between them, then you might notice some engorgement until your breasts get the supply and demand messages sorted out again. This can take a few days. Some fullness is necessary for your breasts to get the message to reduce supply. You can take ibuprofen according to packet instructions, and cool compresses on your breasts for about 10-20 mins at a time can reduce the pain. You might need to express or feed a tiny bit to take the edge off severe discomfort.

If you want to stop breastfeeding all together, replacing a regular breastfeed (say your usual 10.30 pm feed) with formula, then waiting a few days and replacing another feed (say 10.30 am) often results in less discomfort than stopping everything suddenly. Other mums do half a breastfeed and then offer formula each time to reduce the supply and demand signals to their breasts.

If stopping all breastfeeding immediately, it is recommended to express the minimum possible to remain comfortable and avoid engorgement and mastitis. Although not licensed for this purpose, over the counter decongestants can have the side effect of reducing milk supply. Cool compresses can help to reduce any pain from feeling over full. Ibuprofen can be useful to reduce inflammation and paracetamol takes the edge off the pain.

Help, is this mastitis? What can I do?

You might have mastitis if you have a hot, red (in pale skin) tender area of the breast and feel like you also have flu. Feed or pump according to your baby's demand, not more. Use cool compresses, ibuprofen and paracetamol. Do not vigorously massage the breast or apply heat. If it is not resolving within 24 hours, or you are getting worse, or both breasts are affected, contact the GP and consider antibiotics.

What is mastitis?

Mastitis is an inflammation of the breast.

It was traditionally believed that mastitis is caused by an **area of the breast not draining effectively**, so milk stays there (milk stasis). A mother might have noticed a tender area of the breast or a blocked duct before getting mastitis. This can happen for several reasons, including the baby not latching well or having problems sucking, the mother missing feeds or leaving long gaps between them, the baby sleeping for longer than before, tight clothing or a woman's fingers digging into her breast and blocking her ducts. It is thought that milk leaks from the blockage into the surrounding tissue, and the body responds in the same way it would to an infection.

However, as research into mastitis improves, we are noticing that it can be correlated with bacterial imbalances or infections. Bacteria may enter a crack in the nipple (often due to a poor latch and the associated damage). Mastitis has also been linked to changes in the microbes of the breast due to shield and pump use. Sometimes mastitis feels like rotten luck. We often never find out why it has happened.

Mastitis feels different from engorgement (which tends to affect the whole breast and make it feel full or tight all over - see

How can I deal with engorgement).

Have I got mastitis?

As with most infections, if you have one, you know about it. A mother with mastitis will feel unwell and may have flu-like symptoms. She may have a temperature so may be hot to the touch. She might have chills and feel cold and shivery (despite having a fever). She may have an angry, hot, tender, red area of breast which will feel swollen and painful. The redness is not visible in darker skin tones, but the affected area may look darker. Mastitis might cause a burning pain all the time, or it might just hurt to breastfeed or express. Some mothers will have nipple discharge, or notice pus or red streaks in their milk.

Yikes, I have mastitis, what do I need to do?

Don't panic. Start some self-help measures. If you are not feeling better or are feeling worse within a few hours, please call your GP or 111.

Self-help measures

Continue to breastfeed or express as normal. Attempting to drain the breast does not work, it has the opposite effect because it tells your body to make more milk. If feeding is painful or you need some help, please reach out for support.

The worst thing you can do is stop feeding or expressing. There is nothing wrong with the milk, even if you subsequently take antibiotics for mastitis. A tiny amount of antibiotic makes its way into the breastmilk, but this is not considered a problem for your baby.

Warm compresses (a flannel soaked in hot water, or a heat pack designed for the purpose), or a nice warm shower can help milk to flow, so use them 5-10 minutes before a feed or expressing session. After a feed, use cool compresses, ice wrapped in some cloth, or special cool packs to help to reduce the pain.

Avoid tight fitting clothing, including bras. Use hydrogel pads to treat nipple cracks.

Rest, nice food and plenty of soft drinks can help. Ibuprofen (400 mgs three times a day after food) and paracetamol (one to two 500 mg tablets four times a day) can be used according to packet instructions. **Do not take aspirin.**

Vigorous breast massage is not recommended for mastitis. Please avoid damaging this sensitive breast tissue with shaking, kneading or pummelling. You might be able to help excess fluid drain back into the lymphatic system with gentle massage towards the armpit. Many women lie on their back to do this, massaging no harder than they would stroke a cat.

If mastitis is getting worse, or on both sides, or your baby is under two weeks old, please call your GP or 111.

Explain that you have a baby and what is going on for you. Many GPs will prescribe antibiotics and have them ready at a local chemist or even couriered to you. If this is not possible, please call 111.

Antibiotics should start working within 12-24 hours and you should start feeling better. You will most likely be offered a ten-day course. Please keep taking them according to packet instructions even when you feel better.

As ever, this is general advice. Please call your healthcare professionals for detailed support for your own unique situation.

If you are seriously unwell, dizzy, confused, develop nausea, vomiting or diarrhoea, or have slurred speech along with the symptoms of mastitis you need to seek urgent medical attention. These could be signs of sepsis, which may require hospital admission and intravenous antibiotics.

Do I have nipple thrush?
Probably not.

Thrush is over-diagnosed, and many breastfeeding mothers and babies are being treated unnecessarily.

Do you have shooting pains in your breast? Are your nipples pink and sore? Does baby have a white coating on their tongue? If the answer is yes and you go to the GP, you are likely to leave with a prescription of thrush mediation for you both.

Over the years I have met many women being treated for nipple thrush and their babies for oral thrush – but I can only be certain of seeing it once. In this section we will talk about thrush signs and symptoms, and why they are often mixed up by GPs who do not have time to watch a feed.

Causes of sore nipples

Sore nipples can make a mother utterly miserable. If you are struggling with sore nipples, especially if breastfeeding has never been pain free, please get some proper support from a lactation consultant or a breastfeeding group. In nine out of ten cases a professional watching a whole feed can offer guidance, unique to you and your little one, that reduces the pain or even removes it all together.

If your baby has a shallow latch onto the end of your nipple, especially if your nipple comes out looking pinched or white at the tip, it is probably compression that is causing the nipple pain and the shooting sensation as the blood rushes back afterwards. This is called vasospasm. It is especially common when a baby has a tongue restriction.

A small number of women suffer from Raynaud's Syndrome, that causes vasospasm in the nipples as well as other extremities.

This can be very painful. Mothers are advised to keep the nipples warm. Some find expressing less painful than direct breastfeeding, and some need pain relief or other specific medication from the GP.

So, when is the pain caused by thrush?

This picture is from the NHS website and shows oral thrush.

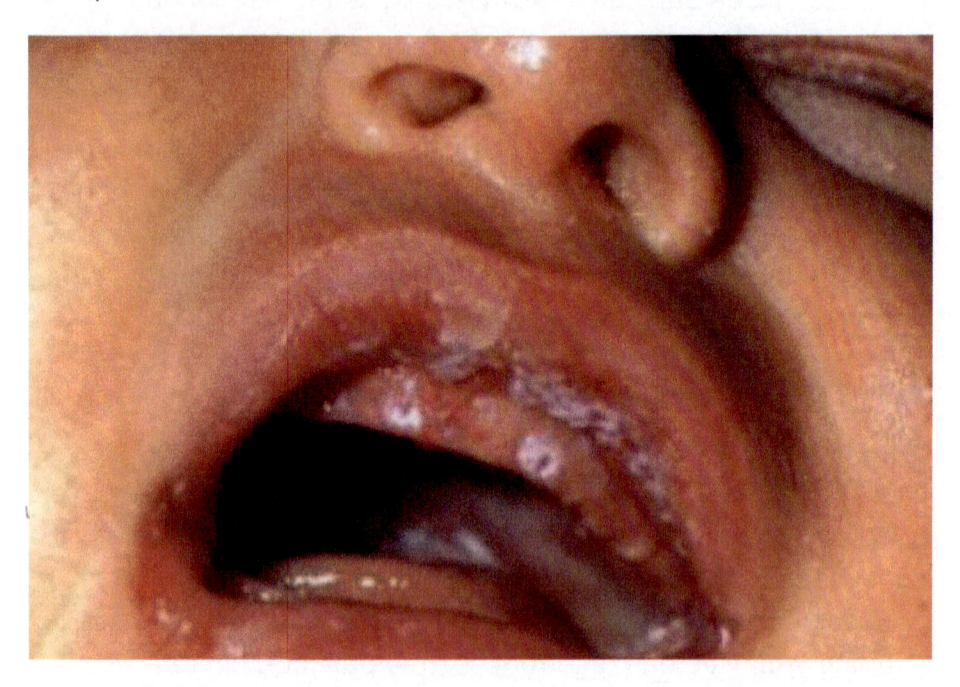

Thrush is a fungal infection of the nipple that can also affect your baby's mouth. Women often describe an itchy, scratchy, broken glass kind of pain in their nipples and shooting pains deep in their breast for up to an hour after feeds. This happens every feed and in **both breasts**.

A baby with thrush will have **irregular white patches** in their mouth, extending across the tongue, cheeks and gums and sometimes lips. The patches do not rub off. Sometimes the tongue and lips have a white sheen. As thrush passes between you, it almost always happens on both breasts at once.

Thrush is uncommon in mothers and babies within the first two weeks, even if the mother has been given antibiotics in labour or after the birth. Early onset nipple pain, especially if breastfeeding has never been pain free, is unlikely to be thrush.

What is that white coating on my baby's tongue then?

Babies can often have a thick white coat on their tongue. It is known "milk tongue."

It occurs in babies whose tongues are a little restricted in their mouth, for example those with tongue-tie. These babies cannot raise their tongue well and rub the top of it during feeds. If it is not hurting you or them, and your baby is happy and gaining weight, please do not worry about it. If you also have nipple pain it might be the tongue restriction that is causing the problem.

As a starting point, please get help from a professional in achieving a good deep comfortable latch. It might be necessary to treat your baby's tongue, and you can find help with this through your medical professionals or the Association of Tongue-tie Practitioners.

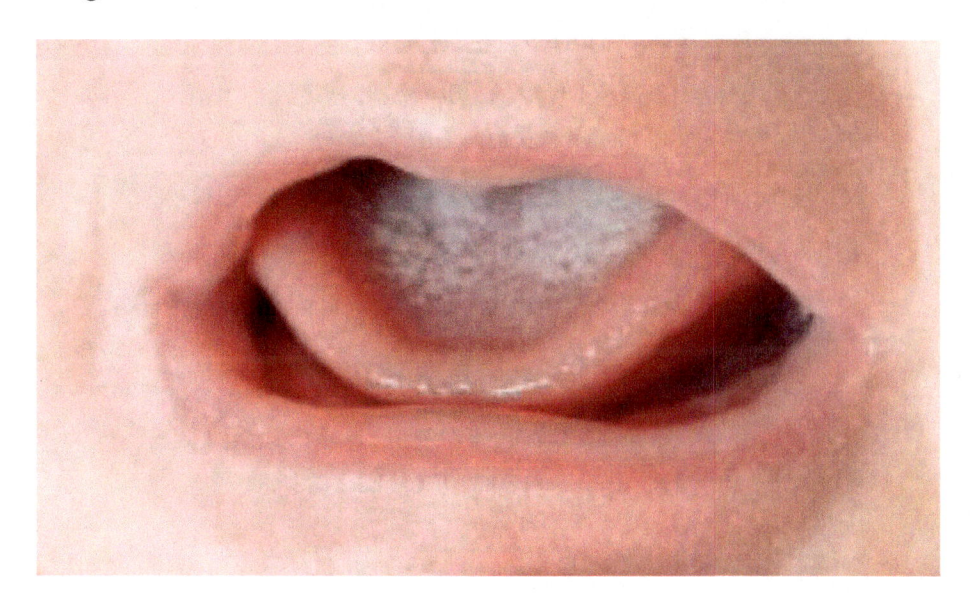

What if it really is thrush?

Please see your GP and **get a swab** to confirm this. If it is thrush then treatment will be prescribed for both you and your baby. It might take a couple of weeks to go away. The evidence-based treatments are listed in an excellent publication from The Breastfeeding Network and summarised here.

Thrush passes between mother and baby, and sometimes the rest of the family. It is essential to consider hygiene, including changing breast pads frequently, washing bras regularly, avoiding sharing towels between family members.

Thrush treatments listed by the Breastfeeding Network

If swabs are positive, treatment will be prescribed for mother and baby. The suggested treatment is:

- Miconazole oral gel applied gently a small amount of time to your baby's mouth four times a day.
- Miconazole 2% cream applied sparingly to mother's nipples after every feed.

Most cases begin to feel better after 2-3 days but thrush takes longer to completely disappear.

If symptoms persist:

- Ongoing topical treatment plus.
- Oral fluconazole tablets 150-400mg as a start dose and 100-200mg daily.
- For nipples that are very inflamed a mild topical steroid cream can be used to facilitate healing. Miconazole 2% plus hydrocortisone 1% cream may be useful.

Anecdotally some mothers find acidophilus capsules can help to restore bacteria which can keep thrush under control (available from health food stores or chemists). Some mothers find reducing the level of sugar and yeast in their diet helps. There is some research about the effect of grapefruit seed extract in inhibiting thrush, so I hope this leads to alternative treatments in the UK.

My nipple has gone white at the tip and feels sore after breastfeeding!

Blanching or vasospasm happens when either the tip or whole nipple has restricted blood flow. The nipple goes white and then sometimes blue. After a time, the blood rushes back and the nipple turns purple-red. The nipple may feel numb, burning or tingling and there can be stabbing pain deep in the breast. Vasospasm can last for several minutes, and the stabbing pains can happen between feeds.

The most common physiological causes for these blanching episodes include:

- Incorrect positioning and latch.
- Baby having a crunchy tight suck and sometimes tongue-tie.
- Mothers with Raynaud's Syndrome.
- Some drugs, such as theophylline, terbutaline, epinephrine, norepinephrine, serotonin, nicotine, and caffeine.
- Mothers who have previously had breast or nipple surgery.

Exposure to cold makes nipple blanching worse, and mothers are advised to keep warm and even use heat packs after nursing.

Any mother with nipple blanching would benefit from a proper assessment of her breastfeeding by a trained lactation consultant to figure out what is going on. Possible treatment measures depend on the cause.

The Breastfeeding After Reduction website suggests that in the case of Raynaud's Syndrome, blanching can sometimes be improved by using food supplements such as calcium and magnesium, as well as evening primrose oil (gamma linoleic acid), and fish oil (eicosapentaenoic acid and docosahexaenoic acid) eicosapentaenoic acid and docosahexaenoic acid). Unfortunately, it can take up to six weeks to see improvement with these supplements.

When the discomfort from nipple blanching is severe, mothers may approach their GP for medication. Many women are prescribed nifedipine, which is a calcium channel blocker. It has been shown to be clinically effective in reducing nipple blanching fifty to ninety-one percent of the time. A small amount of it passes into breastmilk, but this is not considered a significant risk to the baby. As ever, there are some possible side effects, including headache, flushing, dizziness, rapid heartbeat, and oedema in the extremities.

What is a nipple bleb and what can I do about it?

A nipple bleb looks like a little white blister. The breast tissue leading to the bleb can feel swollen and blocked. For something so small, a bleb can be remarkably painful. A mother with a bleb will often have shooting pains in her breast tissue, and find breastfeeding excruciating.

A bleb can form when skin tissue grows over the end of a milk duct, and the duct may then get blocked inside with thickened milk or debris. This prevents the rest of the milk from the associated part of the breast from draining properly, causing pain. Alternatively, it is thought that internal inflammation can cause debris to collect in the milk ducts, which can result in tiny little pebbly particles that need to pass out of the duct and can get stuck.

What causes blebs?

It is not entirely clear what causes blebs. Some mothers seem more prone to them than others. Blebs can result from a suboptimum latch, especially where the nipple skin is getting repeatedly rubbed. They can also be associated with badly fitting shields or pump flanges. There is increasing evidence that they might be related to imbalances in the bacteria of the breast microbiome, exacerbated by antibiotics and the use of shields, pumps and similar products.

How can I treat a bleb?

To clear the bleb, the mother might try various techniques. It is often recommended to keep the bleb moist so that her baby can suck it away. This is harmless to her baby. An effective technique is to soak a cotton wool ball in olive oil and wear it in the bra against the bleb. This can take many hours and sometimes days. Some women use clean, moistened warm muslin cloths or flannels to rub the area. Others try taking a warm bath, or soaking the breast in warm water before a feed. The use of Epsom Salts is no longer recommended.

It can help to roll and pull the nipple gently with the fingers, to loosen up the bleb before your baby latches.

You might find a short course of a lecithin supplement helps over the acute phase of the bleb.

The Academy of Breastfeeding Medicine advises a mild steroid cream to treat a bleb, and not to try to take the top off them.

However, many women find that using a sterile needle to remove the skin covering the breast pore is the quickest and easiest way to relieve the bleb. If doing this, start with clean hands. Use a sterile needle to release the bleb from the side, not pushing the needle into the body of the nipple and not drawing blood.

A GP or practice nurse can do this for you if you prefer. Apply a hydrogel pad or nipple cream to allow for moist wound healing afterwards.

To avoid further blebs, you might like to get a professional to review your baby's latch. Consider improving your microbiome with probiotics, and think about your overall diet, perhaps including more choline. Make sure that any pumps or shields fit well and are not causing nipple trauma.

You might find that you get a few blebs and then they just go away and don't come back.

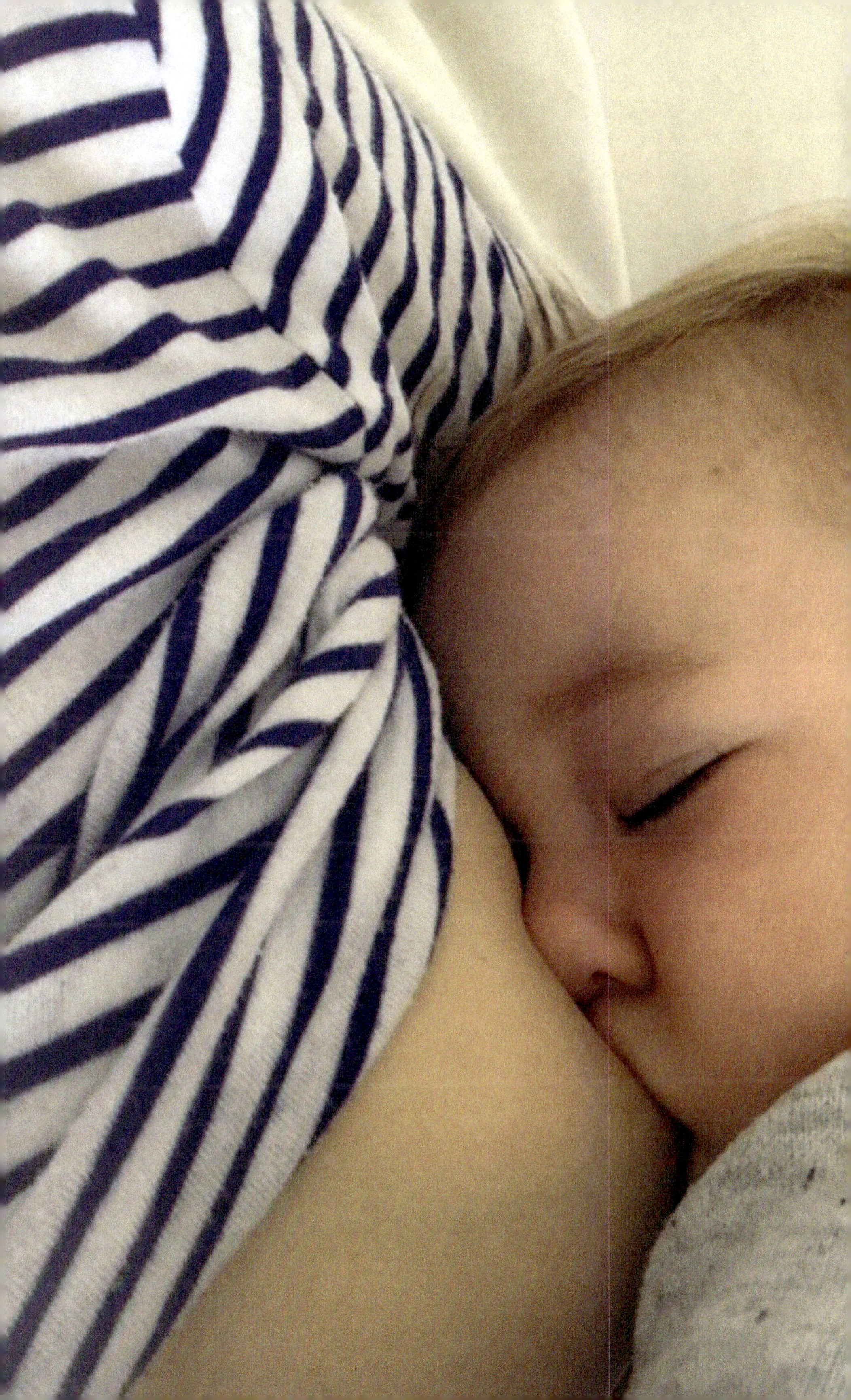

Milk supply issues

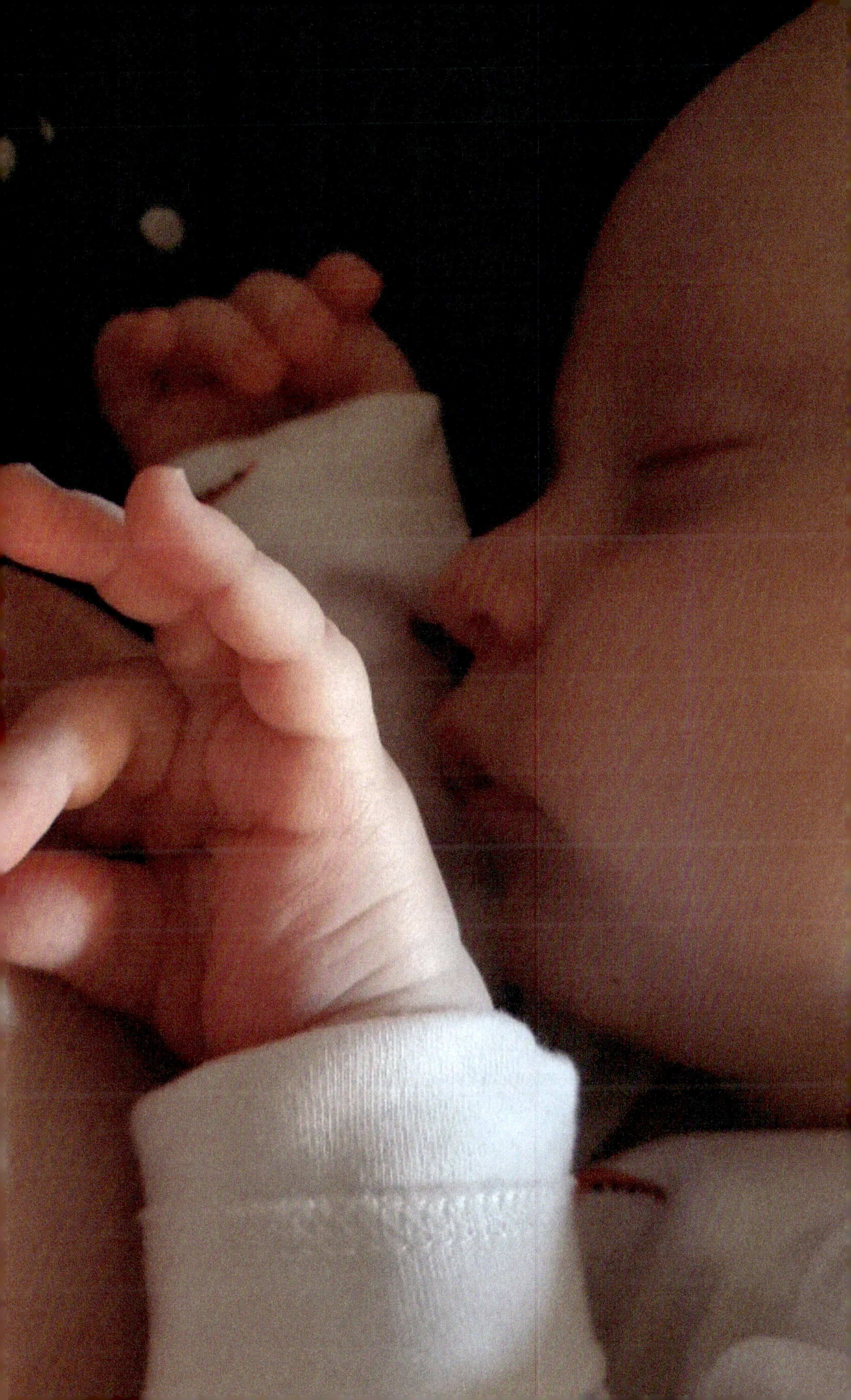

Many new breastfeeding mothers are concerned that their little one is not drinking enough. This section covers what is normal, recaps ways to tell how feeding is going, and what to do if milk supply is an issue for you.

Do I have enough breastmilk for my baby?

Lots of women who worry about their milk supply do not actually have a problem!

We live in a bottle feeding culture, with the idea that babies should be fed at pre-defined intervals until they are so full that they sleep soundly for at least a couple of hours. This is not how things work for breastfeeding mothers, who are often left wondering whether they are doing something wrong when their baby feeds frequently or irregularly.

The bottom line is that babies and breasts are designed to work together in a way that is unique for each family. It is normal and desirable for a new baby to feed frequently, eight to twelve times a day in 24 hours. Breastmilk is easily digestible, and babies are not meant to be asleep for long stretches in the early days.

When a mother feeds responsively, offering the breast when she notices feeding cues, she works with nature. As her baby takes breastmilk the breasts make more, keeping mum and baby in sync. There is no "right" pattern for feeds, and no "one size fits all" for any age of baby. The time between feeds will often space out naturally as babies become bigger and more efficient at feeding, but it is also limited by each woman's body and milk storage capacity. A mother does not have to do anything to try to change baby's feeding pattern, regardless of what her friends, other books or google might be telling her.

If you are worried right now, look at Urgent Signs section of this guide.

How can I tell whether my baby is getting enough breast milk?

The three main questions to ask yourself are:
- How is my baby looking and behaving?
- Is my baby doing appropriate amounts of wees and poos?
- What is my baby's weight?

How is your baby looking and behaving?

You know that feeding is going well when your baby feeds eight to twelve times in every 24-hour period. There will be times of day when the feeds are clustered together, generally in the evening, and times when your baby may sleep for up to three and occasionally four hours. A well-fed baby will be alert and peaceful at some times in the day. A well-fed baby will not be crying all the time or sleeping for many hours in a row.

Normal feeds can be anything from five to 45 minutes, it is what your baby does at the breast that counts. You are looking for active sucking and swallowing, along with the snuggling and pausing that are also natural in a breastfeed. When your baby swallows their chin drops further, there is a tiny pause, and they "huff" out of their nose.

Underfed babies tend to get sleepy rather than scream for food, and we need to wake them if necessary. If a newborn baby is consistently sleeping for more than four to five hours or more at a stretch, they will not have sufficient feeds in 24 hours. Be particularly watchful of a baby that is premature, underweight, unwell or jaundiced. Sometimes these babies do need a feeding regime, often being woken for food every three hours, to keep their energy up until they get better. A very sleepy baby who is hard to wake could be dehydrated and needs urgent medical attention.

If there is a problem, please call your midwife, 111 or even 999 if necessary. Do not sit at home being anxious. Health professionals are very keen to hear from you.

Remember to keep track of wees and poos

The best indication of what is going in, is what is coming out! Some parents use a notebook or phone app to track poos and wees in the early days. Please see the section What should I expect in my newborn baby's nappies? If you are really worried, recap What feeding signs should I worry about in my newborn? and Urgent signs.

Remember that by about six weeks after birth, many babies stop pooing every day and can go several days between poos. This alarms many parents, but it is normal. If the poos are still runny when they come, you probably don't have to worry about it. As ever, GPs and health visitors can chat through any concerns you have about your little one.

What is happening with your baby's weight?

Your baby's weight is monitored closely for the first few weeks. Your baby will be weighed at birth, day five, day 10, and then at intervals to be advised by your midwife or health visitor.

It is considered normal for a new baby to lose up to 8-10% of their birth weight in the first week. Any more than this could indicate an underlying feeding problem and your midwife will advise you on what to do next. Your baby is expected to be back to birth weight by the end of the second to third week. Then most babies put on 150-200 g or 5-7 oz a week.

Am I worrying unnecessarily about my milk supply?

There are some things that you might not have to worry about.

Your baby wants to be held all the time

This is often totally normal. The instant you put your sleepy baby down they seem to wake up and root around to go back the breast. If they are growing well and doing the appropriate amount of wet and dirty nappies, they are clever babies who know it is nice and safe to be held.

Your baby wants to feed frequently

Breastmilk is very easily digested, and your baby loves to be at the breast for comfort and pain relief as well as food. Newborns will often feed every 90 minutes to two and a half hours. We are used to a bottle feeding culture where babies are regularly packed full of formula and then sleep like us after a huge lunch. Massive feeds on a set routine are not best practice.

Your baby feeds for a long time

Normal feeds can be anything from 5-45 minutes. Some feeds seem to roll into a marathon session.

Your baby is fussy at points in the day

New babies fuss, some more than others. There are many possible reasons for fussiness, not all of them are to do with baby being hungry. You might read the section on determining whether gas, reflux, intolerance or allergy are contributing factors.

Your baby is fussy in the evenings

Almost all babies are fussy in the evenings until they are at least 12 weeks old and sometimes more. Most new babies are unwilling to sleep in a cot, and will only settle reliably in an adult's arms.

You can't express much milk

There are many reasons why a mother might not express much milk. When did you last feed? If you are happily in sync with your baby, there is unlikely to be a lot of extra supply to express. Some mums who happily exclusively breastfeed are never able to express a drop.

This sounds obvious, but check your expressing equipment is working, fits well, and is on the right setting!

Breast pumps do not work like babies, and while pumps can be great, we do not love them.

Your breasts do not feel full

Many women never notice fullness or feel their let down of milk. Even if you frequently felt fuller at the beginning of your breastfeeding journey, after a few weeks the supply and demand evens out and you might not feel full between feeds. It is highly unlikely for supply to suddenly decrease for no reason, although some medications or health conditions can have an impact.

Your baby happily guzzles a bottle of formula after a feed

Babies cannot help taking milk from a bottle, especially if it is flowing fast and they are not given the opportunity to stop. Do check whether they need the additional milk you are offering, perhaps by not giving it sometimes, or making sure that the bottle feeds allow your baby the chance to stop feeding.

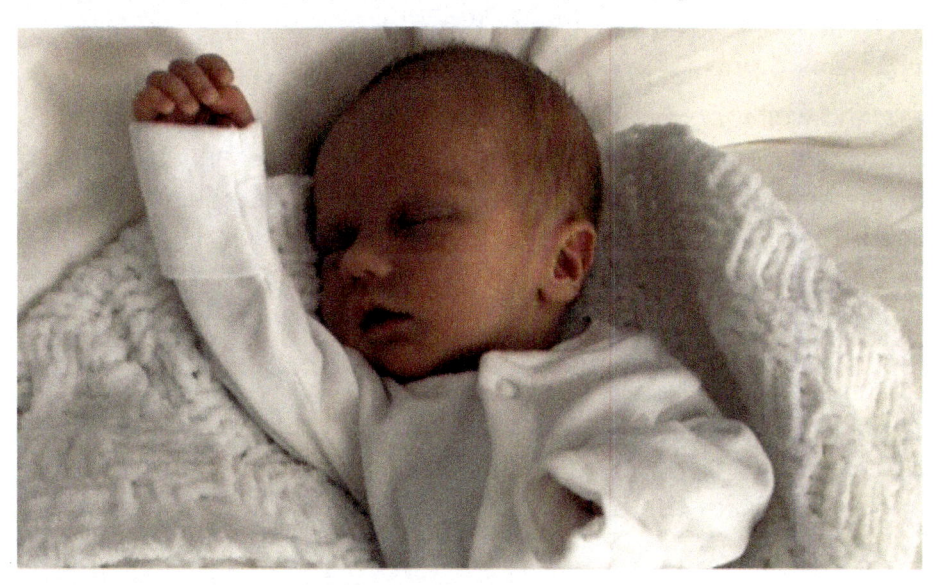

What could cause low milk supply?

Milk supply is complicated. A mother's body, breasts and her baby need to play their part.

How well is the milk factory working?

It can help to think about the breasts as a milk factory; each part of the system needs to work well to make enough milk for your baby.

The controls of the milk factory are a woman's hormones. Prolactin and oxytocin are crucial, and many more besides. A woman's hormonal control can be affected if she has thyroid problems, pituitary issues, diabetes, polycystic ovaries or other conditions.

Unexplained infertility requiring IVF can give us a clue that something hormonal might be happening. Lactation consultants ask about these as part of a medical history.

A larger than average bleed (over one litre) at the time of the birth can often have a big impact on the pituitary gland and affect the amount of milk that the mother makes. Sometimes this is temporary but at other times permanent.

The milk factory itself, the breast tissue and nerves, needs to be working well for optimal milk production. Breast or spinal injuries, conditions or surgeries, even years ago, can have an impact on the complex collection of ducts, nerves and glandular tissue that are needed to make milk. A small number of women do not grow normal breasts in puberty and often do not have the tissue required to fully breastfeed a baby. Mothers in their 40s often report a lower supply than those in their 20s or 30s.

The milk factory's feedback mechanism is supply and demand. **Breasts need to be used frequently so that they "order" enough milk for the baby**.

Milk production will go down if breastfeeding is not frequent. This can happen when:

- Mother and baby are separated.
- Baby can't feed effectively.
- Baby is being fed on a strict routine rather than when showing hunger cues.
- Baby is only being offered one side per feed regardless of hunger.
- Baby is offered formula in place of breastfeeds.
- Mother chooses not to breastfeed for long periods such as overnight.

As the breasts fill up with milk, a chemical inhibiter stops more production so that they don't carry on until they explode! If breasts are being left full the inhibitors are literally telling the body not to make more milk. There is no benefit of spacing feeds so that the breasts fill up.

As babies grow they go through periods of feeding much more than normal, this is nature's way of boosting milk supply for them. Offering formula at this point would stop the "more milk needed" message getting through to the breasts.

Breastmilk production is often most sensitive in the early days and first few weeks after birth. After this, the body gets into the swing of things. Sometimes a lack of effective breastfeeding or expressing in the early days can have a permanent impact on milk supply for that child. It seems that some women's supply is more sensitive than others, it is an under-researched area. If this applies to you, please consider contacting a lactation professional for support.

Some medicines and substances such as nicotine, certain foods, a restricted or low-calorie diet can reduce milk supply. Ironically, drinking far too much water can reduce supply too – just drink to thirst rather than way beyond it.

Can the baby take the milk?

If we are sure that there is milk there for the baby to drink, we need to think about whether the baby can feed well at the breast.

Is the baby swallowing milk? Babies love sucking, so we need to watch to make sure that they are swallowing too. When they swallow their jaw drops and there is a little pause, and they make a tiny huff from their nose. We might even hear them gulp.

A tiny, premature, unwell or jaundiced baby might not have the capacity or energy to breastfeed effectively for a while. For these little ones there is little point them spending energy and getting upset repeatedly trying to latch if they can't do it. They need expressed or formula milk from a bottle, nursing system or cup until they are big and strong and well enough to feed directly at the breast.

In these cases, a mother can protect her milk supply by expressing milk frequently. Lots of baby cuddles skin to skin can help to settle both mother and baby. Often a baby can feed in this relaxed and laid-back position, and it is much nicer than using hands to push baby onto the nipple.

It is possible that a baby is healthy but unable to get milk from the breast because of **problems with achieving a comfortable or effective latch**.

A good latch should not cause the mum pain or nipple damage. If the baby has a latch without much breast in their mouth, or they are tugging the nipple away from where it belongs, they will not be well fed. Laid-back feeding positions help in a lot of cases, as can "nipple sandwich" shaping techniques for flatter nipples and sometimes nipple shields. Long term use of shields is also implicated in low supply, so women should keep an eye on this.

Some babies have **birth injuries or physical problems** that make it harder for them to feed. Some, like instrumental birth bruises or sore necks, will go with time. Others need more investigation. Tongue-tied babies can face challenges with breastfeeding, and may need to be properly assessed by a specialist practitioner.

If you are worried that your baby is not doing well, please seek specialist support.

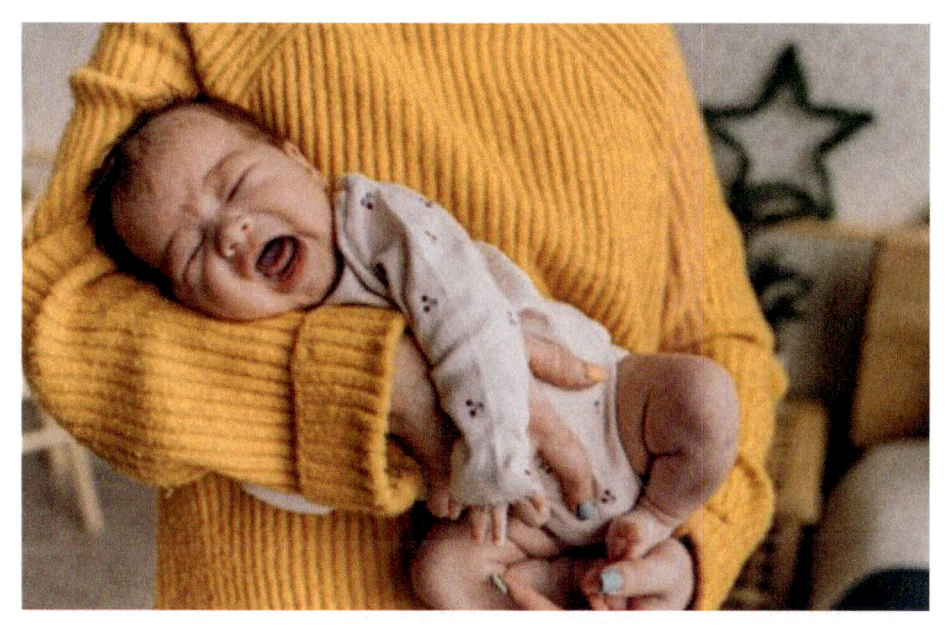

How do I boost my milk supply?

"Demand" more milk.

As breastfeeding works by **supply and demand**, to have more milk we need to increase the demand! This means regularly taking milk out of the breasts by baby feeding and/or expressing. To maximise supply, try to feed and/or express at least eight times a day including at least once at night. It is important not to leave gaps of more than three to four hours between feeds or expressing while boosting supply. Lots of skin to skin contact and unrestricted feeding makes a big difference, and it is helpful if mum gets plenty of support with everything else.

Building up milk supply can take considerable effort and time. Many women will notice a difference after a few days, for others it is a commitment of weeks. Some women may continue to need to supplement with formula.

There are huge emotional and practical implications in working to build up supply, and a woman's support network can make all the difference.

Look out for substances that decrease supply

Occasionally a mother might unwittingly reduce milk supply due to something that she is consuming.

Hormonal medications such as those in the contraceptive pill or some brands of IUD can reduce supply. Several medications inhibit milk production, including some medicines for cold and allergy relief. Always check any medications before you take them, and talk to your pharmacist if you are worried. The Breastfeeding Network has a series of well researched fact sheets about different medications.

Cannabis, alcohol and cigarettes are all knows to reduce milk supply. Nicotine in any form reduces prolactin and oxytocin, key hormones in producing and letting down milk.

Some herbs, if consumed in excess, have been implicated in reducing supply. These include parsley, peppermint, jasmine flower, oregano oil and sage.

Many mothers who are pregnant while nursing experience a drop in supply.

Eat a balanced diet, drink to thirst

Some mothers are worried about their weight after their baby is born, but breastfeeding burns a lot of calories. We know that excessive dieting, consuming less than 1500-1800 calories a day on a regular basis, can reduce milk production. If you do wish to lose weight, aim for slower and more gradual weight loss.

A balanced diet is important in getting all the nutrients required for optimal milk production. This can be hard for a new mum, and some find that they like to take vitamin supplements. Vegetarians and vegans may not have sufficient B12, and this can cause drowsiness in the baby and slowing down of milk production. In this case some lactation consultants suggest having B12 levels professionally checked. Vegetarian and vegan diets may also be low in protein, calcium, iron or healthy fats, so do make sure that you are getting these. Please seek specialist help if you suffer with anorexia or have had gastric bypass surgery.

Please don't worry too much about your fluid intake, just drink when you are thirsty. Drinking too much water, beyond thirst, can reduce supply.

Drugs and galactagogues

Certain medications and remedies are available to boost milk supply, although none will be as effective as appropriate levels of feeding and/expressing. The **Breastfeeding Network** has some well researched information on the medicines Domperidone and Metoclopramide, which are licensed for other conditions but have a side effect of improving milk production in some women. Of the two, evidence suggests that Domperidone is better tolerated with fewer complaints of side effects.

Galactagogues are substances that aid the initiation and maintenance of milk supply at a level which meets the needs of the baby. Almost every culture in the world has a tradition of particular food and drink that are given to nursing mothers. There is a comprehensive collection of information on galactagogues in "Making More Milk" 2nd Ed by Marasco and West. The website www.kellymom.com also lists various milk promoting substances and the best available evidence of their effective doses. It is important to determine the strength and quality of the product you are considering.

Remember that your baby loves you

A mother's ability to breastfeed her baby can profoundly influence her emotions and feelings about herself. Many mothers tell me about their sense of frustration, shame, loss and grief when they cannot fully feed their baby. Some admit to being envious of friends and family who are finding it easy. Some feel that their overwhelming desire to breastfeed against the odds is somehow selfish. Some women fret about decisions they made in the past that are affecting milk supply now. There are so many reasons why breastfeeding might not work out, the majority of which are not fully in the control of the mother. In addition, many mothers do not receive the support or accurate information that they need.

Many women interpret their baby's fussy behaviour at the breast as a sign that their baby is rejecting them or their milk. If this is you, it is not true! It is an interpretation of your baby's behaviour, but it is not how your baby feels about you. You are the person who knows your baby best. Your baby's need and love for you are unchanging even when your baby gets frustrated by the milk factory.

If you are struggling with low milk supply, please be kind to yourself. Being a new mother is hard. Take things one day at a time. Celebrate any breastmilk that you give your baby, and then try to enjoy your time together.

You and your baby are forming an ongoing loving relationship that has little to do with the way they get their milk.

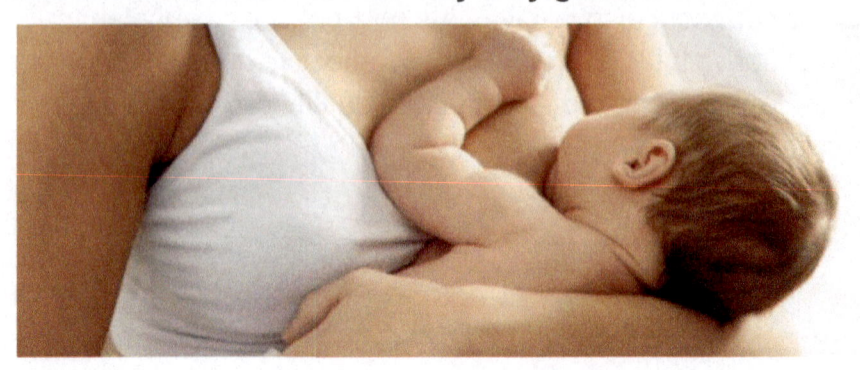

How can I cope with breastfeeding, expressing and topping up?

This is a huge investment of time and energy and not a long term solution! Please get some skilled support.

When a newborn baby is not putting on as much weight as expected, or has lost more than about eight percent of birth weight by day five, mothers are often advised to "top up" after feeds. A typical pattern is to spend a set amount of time attempting to breastfeed and then offer the baby additional expressed or formula milk from a bottle, while the mum is asked to express to protect or boost her milk supply. This can be a brilliant way to get a baby feeding and putting on weight.

The drawback is that parents do not know how or when to stop, and how to wean their baby off the top ups. It is also relentless and exhausting, especially for a new mother who had a rough birth experience and needs time to rest to recover. She will need plenty of caring support around her, as all she can do is focus on this process. It is essential to enlist other trusted adults to manage the coordination of food, work and other caring duties.

It is important to unpick why this situation has happened, and there is a lot more on short and longer term causes of low supply elsewhere in this guide.

How can I stop the cycle of top ups?

Be guided by your baby. If your little one latches comfortably, clearly sucks and swallows milk, gets offered both sides, comes off looking relaxed and sleeps for a bit before waking up hungry, then you might well have cracked it!

Look for wet and dirty nappies appropriate for the age of your baby. By the end of the first week, at least six heavy wet nappies and at least four poos a day would be a sensible guideline.

Even if everything is now looking great, I rarely advise new parents to ditch all the top ups at once. If your baby has been having considerable top ups, it would be unreasonable to expect the breasts to generate sufficient milk immediately. Pay close attention to your little one. There might be several feeds in a day when your baby seems content afterwards. This is most likely in the midnight to midday stretch. Great, perhaps do not offer a top up then. Monitor their behaviour and their nappies. Be prepared to offer the breast responsively, when baby asks for it, even if that is sooner than you expect. If your baby is clearly still hungry and will not settle regardless of what you do, they still require their top up for now.

Continue to boost supply with frequent expressing especially if your baby has not had a good breastfeed at a particular time of day. Offer the expressed milk as a top up.

Parents who have had a worrying time with their little one might still appreciate the certainty of some bottle feeds each day until their confidence has improved. This is totally understandable. Softly does it. With time things tend to improve. And if they do not, and your baby still needs top ups to be sufficiently fed, please try not to blame yourself or them. Sometimes these things just happen. If you have done everything you can to investigate and change the situation and it still isn't working, accept that your baby needs extra milk and they will still thrive in your loving care.

I have too much milk! What can I do?

Consider feeding positions that work against gravity (laid-back or koala). Keep baby's head above their bottom during feeds. Do not encourage more milk with extra expressing. It often helps to offer just one side per feed, but get some skilled guidance if you can.

You would think it would be a nice problem to have. Loads of lovely milk for your baby, rather than struggling to get the last drop with a baby who is never satisfied. To be honest, mums that over produce milk or have a forceful flow can often dread breastfeeding. Their feeds are messy, baby unsettled, and their boobs can get painfully full and lumpy between feeds.

"Oh my goodness, this is a nightmare. My baby cries at the breast and there is milk spurting everywhere. If he comes off coughing it sprays all over the place. He is constantly gassy and unhappy. I can't leave the house, there is no way I could feed him in public."

For most women this problem only lasts a few weeks, although I have worked with mothers where it persists beyond this.

Feeding positions for fast flow or oversupply

Most mums chose laid-back positions that work against gravity and stop the baby from being overwhelmed with milk. The baby may prefer feeding with their head higher than their bottom rather than horizontally across mum. Alternatively, mothers can feed while lying on their side and their baby will typically allow excess milk to dribble from the side of their mouth. I recommend putting a muslin cloth underneath the baby in these instances.

If milk sprays out forcefully at the start of a feed, it can help to let the initial fast flow drain into a muslin or container before attempting to latch the baby. The same goal can be achieved by expressing a little before feeds. Nipple shields can also sometimes help to regulate the flow and make feeding easier for the baby.

Babies coping with fast flow are often a bit gassy and might need more winding than others. If your baby struggles with consistently green poo and explosive nappies, it might be worth talking to your midwife, health visitor or GP.

How can I reduce oversupply?

Generally, supply and demand get into sync over a few weeks with no special action from a mother, apart from feeding her baby when she sees feeding cues.

Oversupply can be challenging. Please remember that breasts work by supplying milk every time they are empty. It can be tempting to pump to relieve the pressure of full breasts, but a lot of expressing increases demand and makes the problems worse.

In cases of severe oversupply, it might help to leave the breasts a little fuller for longer so that they make a less milk over time. Cool compresses, ibuprofen and paracetamol can help to reduce discomfort, but if the breasts are unbearable it can be necessary to feed/express a bit to avoid getting very sore.

Mothers trying to actively down-regulate supply are often advised to block feed. This entails only using one breast for a certain number of hours (usually three or four) before offering only the other one for a certain number of hours. Block feeding can help to send the message to the body that all this milk is not appreciated. I would suggest only doing this with the support of a lactation professional. It is important to bring down supply comfortably and be alert to blocked ducts or the symptoms of mastitis.

Some women are comfortable with producing more milk than their babies need. They might choose to donate it to a milk bank or build up a stash of frozen milk in the freezer.

How can I stop breastfeeding?

Mothers may wish to stop breastfeeding in one go, or gradually wind down to their own or their baby's timetable. As ever, there is no right or wrong way to do it.

Choosing never to offer breastmilk

If a woman never wants to breastfeed and starts with formula straight away, her breasts are likely to still receive the signal from her body to make milk. This can leave her feeling uncomfortable for several days.

Breasts work by supply and demand. If there is no demand for the milk the breasts have receptors within them that get the message not to make more. However, this can leave a woman feeling full and sore. It is no longer recommended to take medicines to "dry up" the milk. Ibuprofen or paracetamol and cold compresses can help with the pain. Some women need to express a little milk to remain comfortable and avoid engorgement and/or mastitis. If a woman is unable or unwilling to do this, she will need to look out for signs of mastitis and get antibiotics if necessary. Although not marketed for this purpose, the over the counter decongestants are sometimes used to bring down milk supply.

When winding down from exclusive breastfeeding

As babies get older they may drop a feed or two, but they still receive a significant portion of their nutrition from breastmilk all the way through their first year. All around the world, some babies continue to take breastmilk through toddlerhood, stopping whenever mum and baby feel like it is right.

Of course, many mothers chose to stop breastfeeding for some or all feeds, way sooner than this.

There are no hard and fast rules on stopping breastfeeding.

Most women find they can comfortably drop one feed every few days with only minimal discomfort. The body soon gets the message that less milk is required.

If a woman gets uncomfortably full she could express a little milk, bearing in mind that she wants her body to receive the message to make less milk overall. If she is keen to keep some milk supply and avoid feeling full, it is wisest to drop feeds at different times of the 24 hour cycle, so that the breasts do not have to spend many hours in a row not being used.

Some women need to breastfeed or express a certain number of times a day to maintain any milk supply. Others have no difficulty and continue maybe with one or two breastfeeds a day for as long as they like.

If choosing, or needing, to stop breastfeeding at once, it is prudent to consider expressing to comfort to avoid engorgement or mastitis. The mother could use the expressed milk for supplementary feeds or cooking for her baby. Obviously do not use the expressed milk if the mother needs medical treatment that is incompatible with breastfeeding.

The emotional fall out

It might be that you are totally ready to reduce or stop your breastfeeding, or it could be something that you do unwillingly. In the UK we hear that most mothers seem to stop breastfeeding before they want to, for a huge array of reasons.

Please be prepared to have some feelings about stopping breastfeeding, which might include sadness, missing the closeness of feeding, guilt or remorse. Some women struggle a great deal with the hormonal changes that come with dropping feeds or ceasing breastfeeding. The hormonal impact of reduced prolactin and oxytocin, coupled with motherhood, can affect women in different and profound ways. A minority of women feel low, anxious or depressed. Please talk to your GP and other trusted professionals if you feel this way.

Let's hope that instead you feel renewed, invigorated, better rested, and able to enjoy an outing with friends while someone else looks after your little one. A wise woman once told me to "*stop breastfeeding on a good day.*"

What is relactation?

Relactation is the process of rebuilding milk supply after a break from breastfeeding. It is easier if a woman has been doing at least some breastfeeding or expressing during this time.

When a woman stops breastfeeding the breasts gradually change their structure and reduce the glandular tissue that makes milk. Breastfeeding is said to have officially ended forty days after the last removal of breastmilk. Even after forty days some women can squeeze out drops of milk.

It is sometimes possible to boost milk supply from previously low levels, depending on a woman's health and how well her breasts are working.

Breast tissue responds to stimulation, so the process requires lots of feeding or pumping. Many babies will not suck for long on a breast with little or no milk, so most women would use an efficient pump many (at least eight) times a day and night to give the body the message to make more milk. It is a labour intensive and often emotional process. Some women consider medication or galactogogues alongside the pumping.

If you would like to relactate, I recommend getting some professional support from an IBCLC, and you will need a lot of love and encouragement from friends and family. You might consider reading *"Relactation – A Guide to Rebuilding Your Milk Supply"* by Lucy Ruddle, and "Making More Milk" by Lisa Marasco and Diana West. There is a Facebook group UK Relactation, Adoptive and Intended Parents Breastfeeding Support

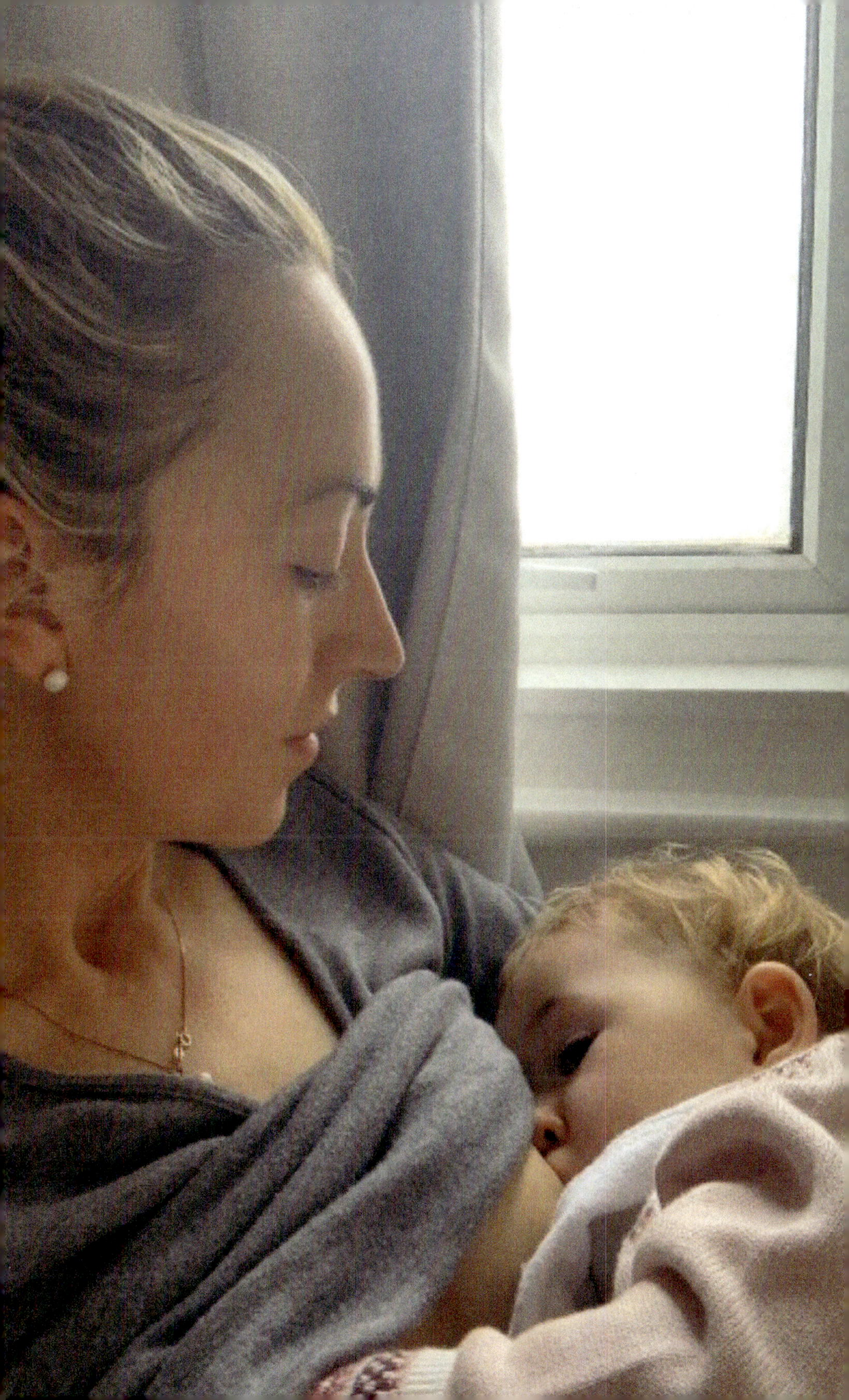

Concerns about your breastfed baby

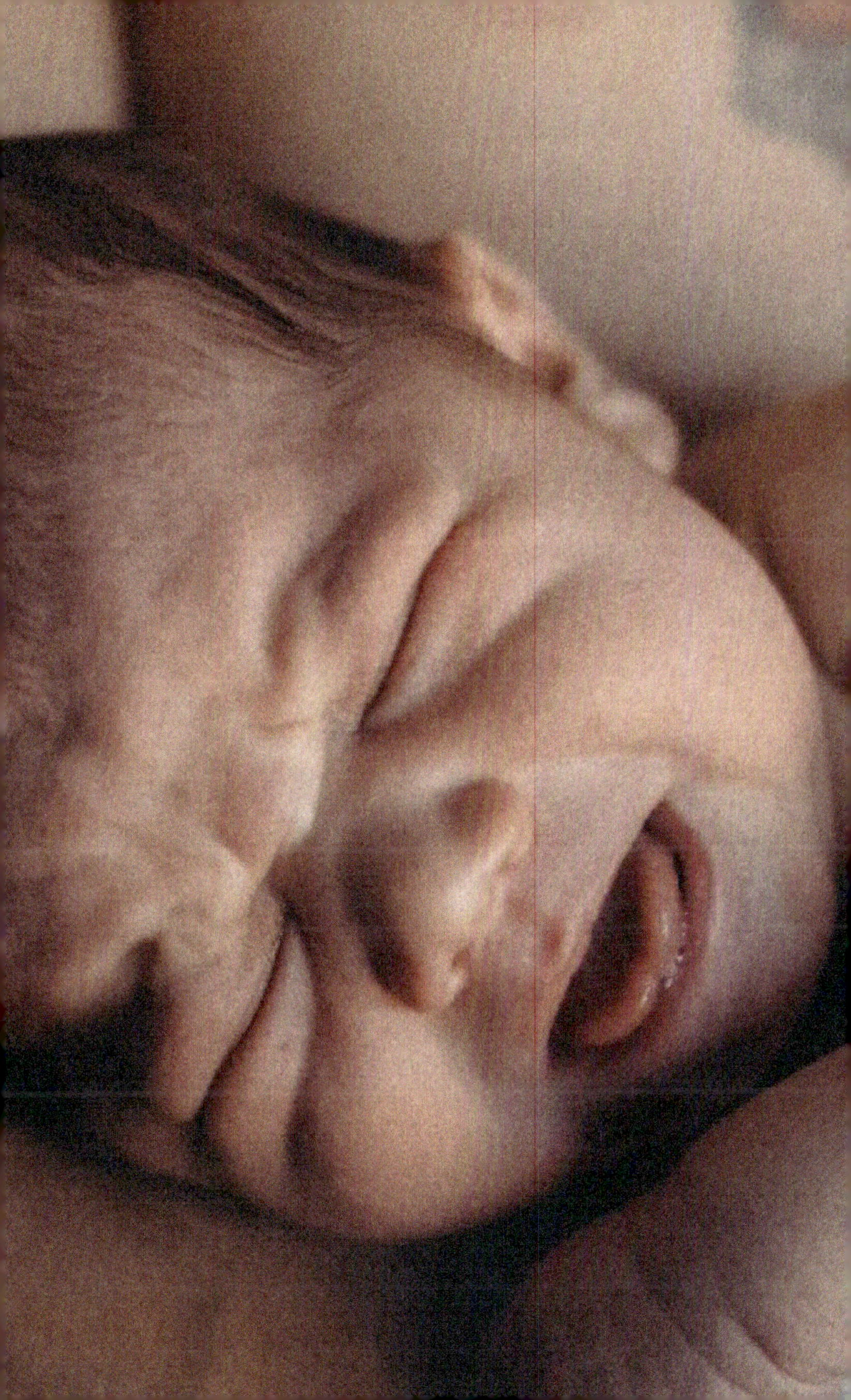

My baby is coughing up mucus, is that normal?

Some newborn babies struggle to clear mucus from their nose and throat. This can be scary for new parents, and their baby can be reluctant to feed. If you see or feel your baby struggling with mucus, if they cry or turn pale or red, quickly lay your baby tummy down on your forearm with their head slightly lowered. Pat your baby's upper back with the palm of your hand to dislodge the mucus and let your baby drool it out.

It is not normal for your baby to stop breathing, go floppy or blue. If this happens, even if your baby seems to recover quickly afterwards, press the emergency alarm in hospital or call 999 for an ambulance and start CPR if necessary.

On the very rare occasions that this happens, I have been surprised that parents tended to think about calling friends or their midwife rather than the emergency services.

Why is my baby spitting up?

Nearly all babies spit up some milk many times a day. This is described by medics as "effortless regurgitation" and commonly known as "posseting," and it does not seem to bother babies very much.

Spit up may look like milk, or a seem bit curdled like cottage cheese. The latter is more common when the spit up happens between feeds. This is all perfectly normal for a baby, as their digestive system is immature and the sphincter to their tummy is a little unpredictable. Your baby might prefer not to be moved around too vigorously immediately after their feeds.

If your baby is mostly happy, feeding well and gaining weight, you generally do not have to worry about their nourishment.

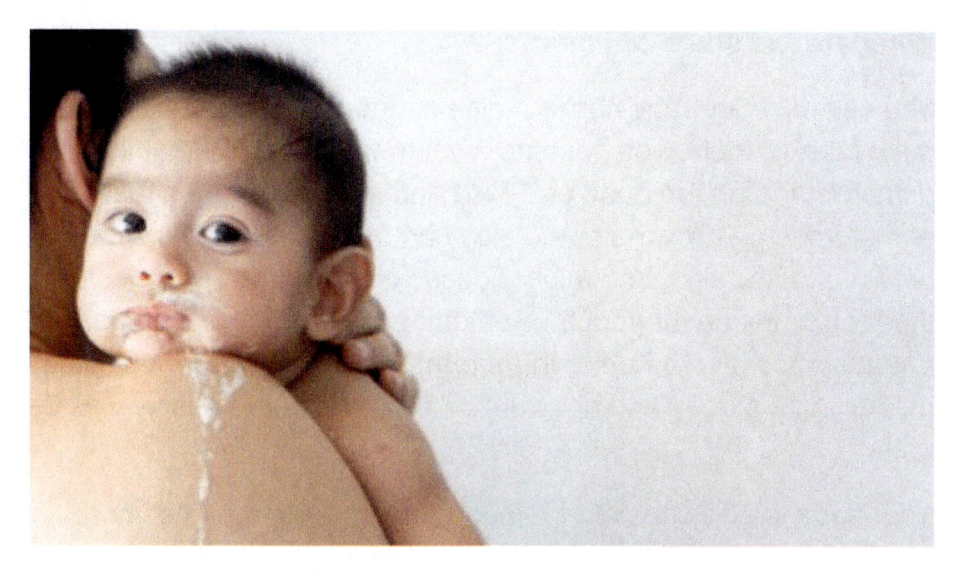

Unlike a posset, which often looks like a mouthful of milk, some babies bring back an entire feed in a gush. About two thirds of babies vomit regularly, with varying force, including some projectile vomits. It slows down after they are about four months old. Of course, if you are worried about your baby, especially if they projectile vomit after almost every feed, then take your baby to the doctor to be checked.

How can I breastfeed my newborn with jaundice?

Jaundiced babies are often sleepy and hard to feed. You might need to wake them every three hours for feeds until they are better, and some require top ups of expressed or formula milk for a while. Severe jaundice needs treatment in hospital, generally phototherapy.

Jaundice is common. Sixty percent of newborn babies become a little jaundiced towards the middle of their first week. This increases to eighty percent of babies born before 37 weeks. Most babies do not require special treatment and it generally goes away by itself within a couple of weeks.

What are the signs of jaundice?

You can tell your baby has jaundice as their skin and sometimes the whites of their eyes become a yellowish colour. The yellowing usually starts on the head and face and spreads to the chest and stomach. Yellowing of the skin can be more difficult to see in brown or black skin. You might notice yellow palms and soles of the feet, or inside the mouth. Sometimes their urine will be darker and their poos will be paler than normal. A newborn with jaundice may also be sleepy and not want to feed, or not feed as well as usual.

Your baby will be checked for jaundice as part of their newborn physical examination. However, if you are already at home, it is important to contact your midwife straight away if your baby's symptoms become worse or if they become very reluctant to feed.

What causes jaundice?

Jaundice is caused when the baby's body breaks down the extra red blood cells that are no longer required, with the liver creating a yellow biproduct called bilirubin. Most babies overcome this jaundice without special treatment, and there is some suggestion that it has a protective effect against sepsis. Jaundice tends to go by 10-14 days as the liver gets better at its job.

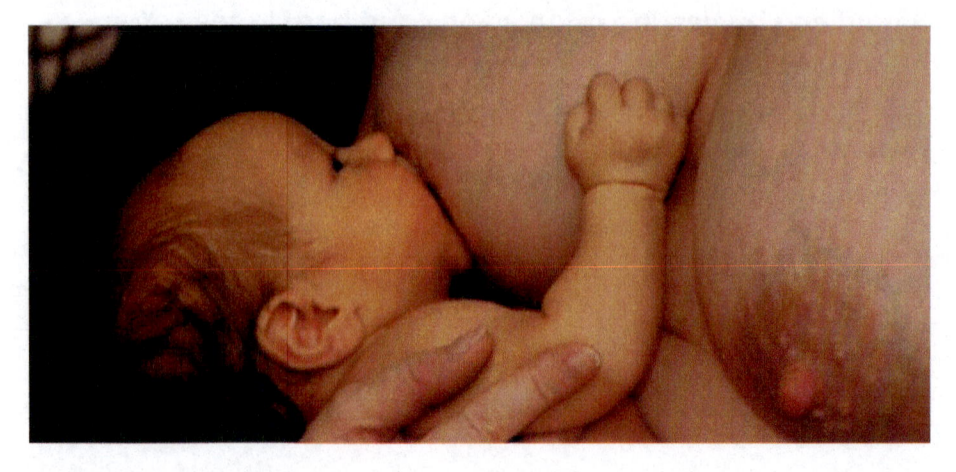

How is jaundice treated?

For most babies, no special treatment is necessary. As with all newborns, it helps if they have frequent feeds. It could be sensible to wake your baby if necessary to feed so that they have milk at least every three hours until they are better. Expose your baby to some indirect sunshine if possible, as this breaks down the bilirubin. Obviously don't let your baby get sunburned.

Unfortunately, some babies develop jaundice more quickly or more seriously than normal. About five percent get jaundiced to a degree where they do need treatment, because of the small risk of the bilirubin passing into the baby's brain and causing damage. Midwives can test for jaundice with a heel prick blood test, or a hand held monitor that is placed against the baby's skin.

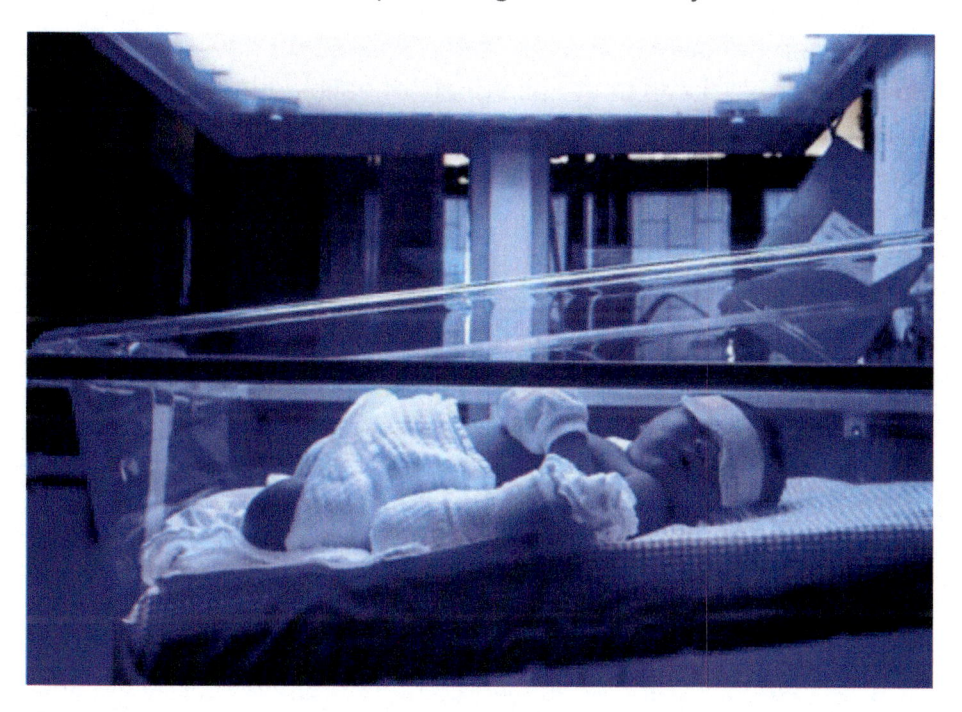

If your baby's bilirubin levels are above the threshold for treatment, they will most likely be given a feeding regime where you feed at specific times and top up with formula if necessary.

They will also be given phototherapy, where they are put under a special light in their cot, a little like a tanning lamp, and will wear soft pads over their eyes to protect them. Some may be wrapped in a more portable biliblanket, that does the same job but allows more cuddles. They will be checked at regular intervals until their bilirubin levels have gone below the "treatment line."

Rarely, a baby will require an exchange transfusion, where the baby's blood is gradually removed using a thin tube placed in their blood vessels and replaced with blood from a matching donor.

The good news is that most babies respond well to treatment and can leave hospital after a few days.

Sometimes jaundice can be caused by another health problem. Causes of pathological jaundice include hypothyroidism, blood group incompatibility or rhesus factor disease, urinary tract infection, or blockage in the bile ducts of gallbladder. In the unlikely case that any of these apply to your baby, your medical team will be able to tell you more.

Why does my baby have green poo?

Green poo is expected on days three and four post birth. Rest assured that the occasional green poo is completely normal. Sometimes it happens if your baby is recovering from a stomach upset.

Some parents worry because their baby does a green poo, and Google suggests that green poo might be a sign of too much foremilk or even lactose intolerance.

Please do not self-diagnose lactose intolerance in your baby. Truly lactose intolerant babies would be very unwell. They would be in severe discomfort with low weight gain, and may have explosive frothy green poos all the time. If you are worried about your baby, please have them reviewed by a medical practitioner.

If your baby is prone to frequent green poo, and you have a very generous milk supply or fast let down, you might be advised to let the first minute of the let down to drain away into a muslin, and to keep your baby on one breast for longer before offering the other. There is more in this guide on switching sides, and managing with too much milk.

Is my baby constipated?

Probably not, unless they are in lots of pain and their poo is hard like pellets when it comes.

Newborn babies seem to poo all the time. Their tiny bodies are set up so adding milk in one end tends to provoke a bowel movement at the other! Over a few weeks this reflex stops, and babies hold onto their poo a bit longer. When babies stop pooing frequently their parents worry that their little one is constipated. But are they?

From about four to six weeks old, it is common for a baby not to poo every day. In fact, they might go several days between poos. If they appear well, and the poo is runny when it finally comes (generally in a poonami) then your baby is fine! This is the case even if baby is wriggly or red in the face and appears to be "straining" when they need to poo. Lots of babies do this, and they are not constipated if the poo soft.

A lack of bowel movements in a newborn baby can be a sign of insufficient milk, so do check with your midwife as a matter of urgency.

Babies struggling with constipation are very uncomfortable, clearly in real pain, and have small hard poos. This is very rare in breastfed babies, slightly more common in formula fed babies, and can happen with those on solid foods. Ask your GP or pharmacist for advice.

To avoid constipation in a formula fed baby, always make sure that you make up formula powder correctly. It is important to add the powder to the water so the concentration is correct.

Breast and formula milk contain lots of water. They are not the cause of constipation. Neither breast or formula fed infants need additional water or juice to drink before they are six months old. Even after six months, breast or formula milk is the best drink for your baby. From a year old, breastmilk remains the best drink for your little one, but formula can be replaced with full fat cow's milk instead.

What is tongue-tie and what shall I do about it?

When the membrane under the tongue is short or tight it is called a tongue-tie, or ankyloglossia.

Many of us will have seen our lingual frenulum, it is the membrane underneath the tongue that links the bottom of the tongue and the floor of the mouth. There are natural variations in the thickness and position of the frenulum, and in some people it is not visible at all.

Many parents, especially those experiencing ongoing breast and bottle feeding difficulties, worry that their baby's tongue might be to blame. It can be a good idea to seek a specialist assessment from someone trained in breastfeeding and tongue-tie to be sure.

The NICE Guidelines in the UK suggest that many tongue-ties do not cause symptoms, do not require treatment and resolve over time. However, some practitioners believe that if a baby with a tongue-tie is having trouble breastfeeding, they should have surgical division of the lingual frenulum as soon as possible.

The NICE Guidelines state *"current evidence suggests that there are no major safety concerns about division of ankyloglossia (tongue-tie) and limited evidence suggests that this procedure can improve breastfeeding. The evidence is adequate to support the use of the procedure."*

How many babies are affected by tongue-tie?

We do not know. While babies are checked for tongue-tie as part of their newborn check, this is not a full or comprehensive assessment of tongue form and function. The process for assessing and grading tongue-ties is not yet universal, with individual practitioners using different methods, so research shows that anything from 3-10% of babies could be affected.

It seems that some babies who appear visually to have a tongue-tie go on to feed perfectly well and never need follow up. Others will not breastfeed well and may be put straight onto a bottle. In either case, we do not always know the outcomes.

What would I look for?

Some tongue-ties are easy to spot. Where the membrane is attached at, or close to the tongue tip, the tongue tip may look blunt, forked or have a heart shaped appearance. These are known as anterior tongue-ties.

The following images show clear anterior ties. The tongue is being held down by the frenulum that attaches right to the tip of the baby's tongue. These babies might have trouble feeding from either breast or bottle unless the tongue-tie is released.

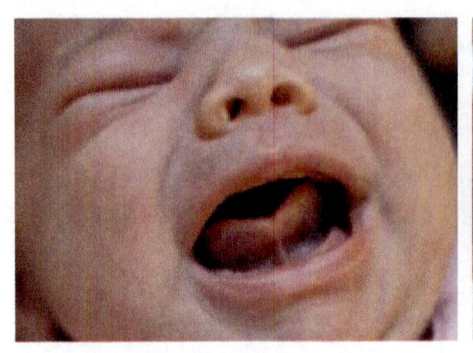

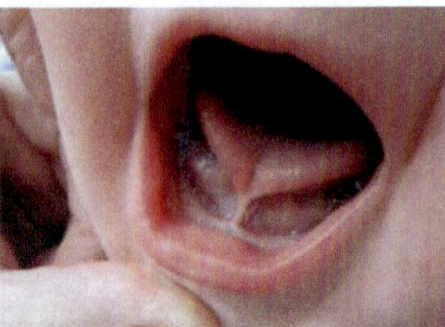

Sometimes the tongue may look normal and yet the baby still has difficulties lifting and using it. They might not be able to lift, extend or lateralise their tongue. When they feed, they might retract their tongue so that their gum touches the nipple instead. These restrictions are called posterior tongue-ties. Not all healthcare professionals believe that posterior ties cause a problem. Many babies with these symptoms seem to benefit from osteopathy or other treatments to release tension in their neck and jaw which can improve tongue function. The baby below may have a posterior tongue-tie.

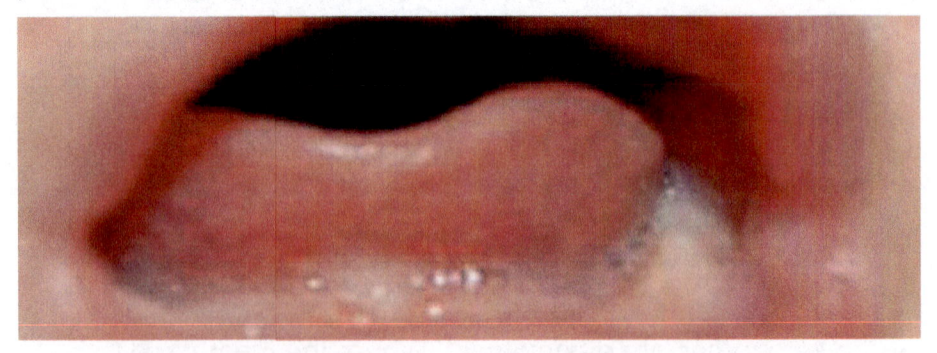

How do I know that a tongue-tie is causing problems?

It can be hard to know, as many other issues can cause breastfeeding problems. Seek proper assessment.

Tongue-tie causes:

- Sore, damaged, bruised nipples.
- Nipples which look misshapen or blanched after feeds.
- The feeling that your baby is crunching or biting your nipple during feeds.
- Blocked milk ducts and/or mastitis without another obvious cause.
- Low milk supply.
- Exhaustion from frequent/constant feeding.
- Distress from failing to establish breastfeeding.

Your tongue-tied baby might experience:

- Restricted tongue movement.
- Small gape, resulting in biting/grinding behaviour.
- Unsettled behaviour during feeds.
- Difficulty staying attached to the breast or bottle.
- Frequent or very long feeds.
- Excessive early weight loss/ poor weight gain/faltering growth.
- Clicking noises and/or dribbling during feeds.
- Colic, wind, hiccups.
- Reflux (vomiting after feeds).

How is a tongue-tie assessment done?

Assessment for tongue-tie requires training and skill and involves placing a finger in the baby's mouth and observing how the baby uses their tongue. It cannot be done by taking a look.

A professional, qualified in tongue-tie, will take a medical history and discuss the birth and your subsequent feeding challenges. They should observe a feed to look at how your baby is coping at the breast or bottle, and consider other possible causes of difficulty.

A tongue-tie assessment is often done with the baby on the assessor's lap or a flat surface such as a table or couch. The assessor will look at whether the baby can lift their tongue past halfway (elevation), move it side to side (lateralisation), and stick their tongue out (extension).

Elevation can most easily be assessed when the baby cries. With the mouth wide open, the tongue tip should lift to at least the mid mouth. In tongue-tied babies the tongue often stays quite flat in the floor of the mouth, or the edges curl up to form a bowl shape or 'v' shape. When the assessor runs their finger along the top ridge of the bottom gum the tip of the baby's tongue should follow the finger so the tongue sweeps side to side. Babies should be able to poke their tongue tip out well over the bottom lip when the bottom lip is stimulated.

Assessors put their finger into the baby's mouth (nail side down) to see how well the baby sucks. They feel for whether the baby is cupping their finger with their tongue, how they move their tongue, and whether they suck the finger in. They also feel the roof of the baby's mouth to get clues from the shape of the palate.

Assessors sweep their finger under the baby's tongue so they can feel the extent of the tongue-tie and will lift the tongue to look. They will note what the frenulum looks like, how thick and stretchy it is, and where it attaches to the tongue and the floor of the mouth.

If my baby has tongue-tie, what should I do about it?

It depends on how severely the tongue-tie is impacting on you and your baby. The NICE Guidelines suggest that babies may benefit from tongue-tie release. In the end, it is your choice. Babies who are struggling with bottle feeding might also benefit from tongue-tie release, but some NHS clinics are reluctant to do it for this reason.

How is the release procedure done?

The procedure can be done in hospital or privately at home or in a clinic. You can find a qualified tongue-tie practitioner at www.tongue-tie.org.

First the practitioner will run through a consent process, talking about the procedure and its pros and cons. If the carers agree to the procedure, the baby will often be taken to another room to have it done.

The baby will be swaddled and supported at the shoulders and head. If the baby cries, it is usually at this stage. The practitioner will lift the tongue and cut the frenulum with sharp, sterile, blunt ended scissors. Some practitioners use a laser.

These pictures show a baby's tongue before and after the release of a posterior tongue-tie.

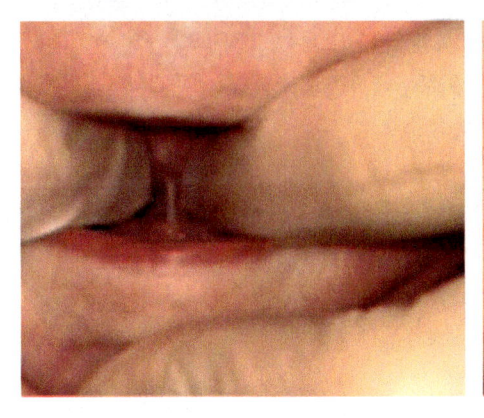

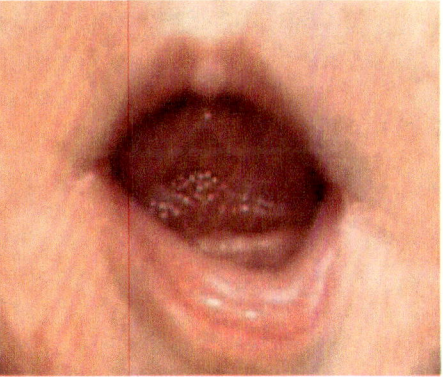

Does it hurt?

In the early days, babies do not seem to require pain relief for the procedure. There are few pain receptors in the lingual frenulum. Many healthcare professionals suggest that it hurts no more than the heel prick test that babies have on day five, or having an ear lobe pierced, and that the pain is momentary. When I have seen it done, the babies seem to object more to being held still than to the procedure itself. Some babies practically sleep through it. It only takes a heartbeat, and then they come back to their mum for a lovely cuddle and pain-relieving breastfeed.

I have known some babies to be uncomfortable for 24 hours after the procedure, and it is estimated that it takes several days for it to heal completely. Releasing posterior ties may be less comfortable than treating thin anterior ones.

A local anaesthetic may be recommended for older babies.

Does it bleed a lot?

Generally, it does not bleed a lot as there are not many blood vessels under the tongue. A piece of sterile gauze may be placed under the tongue to mop up any blood spots while the baby is brought back to their mum for a feed. Mothers are advised to breastfeed immediately to provide comfort and natural pain relief. Breastmilk is healing and antiseptic, and the process of feeding tends to calm the baby and stop any bleeding. Parents may notice a little dark digested blood in their baby's nappies afterwards.

Will it get infected?

Infections are not common. Breastmilk is naturally healing and actively prevents infection.

Will it regrow?

There is a suggestion that some babies' tongues might heal back into a tie again. The organisation of tongue-tie dividers in the UK estimates the chance of this to be less than 3% for anterior ties.

Do I have to do exercises with my baby afterwards?

Some practitioners recommend a series of exercises for parents to do with a baby for several days after the procedure. Many of these are fun.

Other practitioners may suggest massaging under the tongue in a way that is intended to stop the tie from regrowing. This is controversial and many parents find doing the massage deeply upsetting for them and their baby. As I understand it, there are no well designed randomised controlled trials that point to the need for these massage procedures.

Is there a time limit to getting the procedure done?

Many NHS hospitals offer tongue-tie procedures to babies referred between one and six weeks old. After 12 weeks old, some providers insist that a baby receives an anaesthetic for the procedure.

What is the aftercare?

Aside from the exercises that might be suggested, there is little specific aftercare. Babies often develop a white, diamond shaped scar under their tongue.

Some clinics invite parents and babies back for a follow up appointment to assess healing and find out how feeding is going.

Does it help with breastfeeding straight away?

Many women notice an immediate difference in breastfeeding after their baby's tongue-tie release. However, it is fair to say that many do not. People say it can take up to two weeks for the baby to learn to use their tongue again. It is almost like having a newborn. It stands to reason that the tongue is only one element of the feeding equation. A baby's cheek, neck and jaw muscles and the space in their mouth also have a part to play in how well they feed. It is often recommended that parents take their baby to an osteopath trained in this area for some treatment on their muscles and joints alongside the tongue-tie release.

What if we don't do it? Will it be hard for our child in the future?

Nobody can say. Some babies appear to grow out of their tie as they get bigger. The literature is not always in agreement on the longer term impact of tongue-ties, and everyone is different. Older children and adults with anterior tongue-tie may sometimes develop speech impediments, have trouble playing some musical instruments, may not be able to French kiss, and might have pain such as headaches. Many have no trouble at all.

Is lip tie a problem?

A lip tie is a condition where the skin behind the upper lip is attached to the gums in a way that can prevent full lip movement. Currently there is little published research into how lip ties cause breastfeeding issues, or the best way to resolve the situation. Release of lip ties is not recommended for feeding problems. Some dentists release lip ties when a person develops an associated gap between their front teeth.

Why is my baby fussy?

It could be any number of things, many of them quite normal, and fussy babies tend to get better in time. This section offers some trouble shooting.

When babies are upset it is miserable for both the babies and their caregivers. Mothers are heart-broken at the distress of their precious babies, try everything to resolve the problem and feel like they are doing something wrong. Their babies often fuss and arch back during feeds, they may be clingy and impossible to settle, they want to feed and then seem agitated and upset when offered milk, their tummies seem to cause them pain, they are gassy and may bring back milk. What is going on?

After all these years working with new mothers and babies, I know you are not going to want to hear this. The truth is, we often do not know. We really don't. We can have some ideas that might or might not be useful in your situation. We can help to resolve breast and bottle feeding issues and we can certainly test for some of the less common medical conditions that might be causing the crying and fussing. However, in many cases there is no medical consensus about the causes of, or treatment for, these symptoms. Most babies grow out of it within about 16 weeks. Some take longer. In fact, the first 16 weeks is sometimes known as "The Crying Period." It is vital that new families feel supported at this time, because it is horrible, makes everyone tired and anxious, and can have a huge impact on relationships.

As adults we are used to a world of cause and effect. We control the inputs and get the prescribed outcomes. Babies that cry and fuss do not conform to rules that we can explain. There is precious little reliable research evidence to draw on. There are just too many variables. The actions or treatments we try might or might not help for a while. It is hard to tell because many people use several interventions at once, the placebo effect is significant, and most babies grow out of it anyway.

If you are in this situation, here are common sense things to think about.

Is my baby hungry?

The first thing to rule out is hunger.

Is your baby doing an appropriate number of wet and dirty nappies? Are they ever settled? How are they sleeping? Your baby will cry and fuss if there are unresolved breast or bottle feeding issues. These could be due to many things including your baby's position for feeding, the way they are able to use their mouth and tongue, the available milk supply and speed of let down.

I recommend seeking skilled advice and having a professional watch a breast or bottle feed before embarking on any other forms of treatment. In many cases undiagnosed tongue-tie or a super-fast let down contribute to the problem.

Is my baby in need of stimulation?

Does your baby have a high need for cuddles, physical comfort, rocking and other vestibular stimulation?

Many babies benefit from being carried close to their carer in well-fitted slings. Well-meaning advice about not "overstimulating" your baby is counterproductive for most little ones. Babies need lots of loving cuddles.

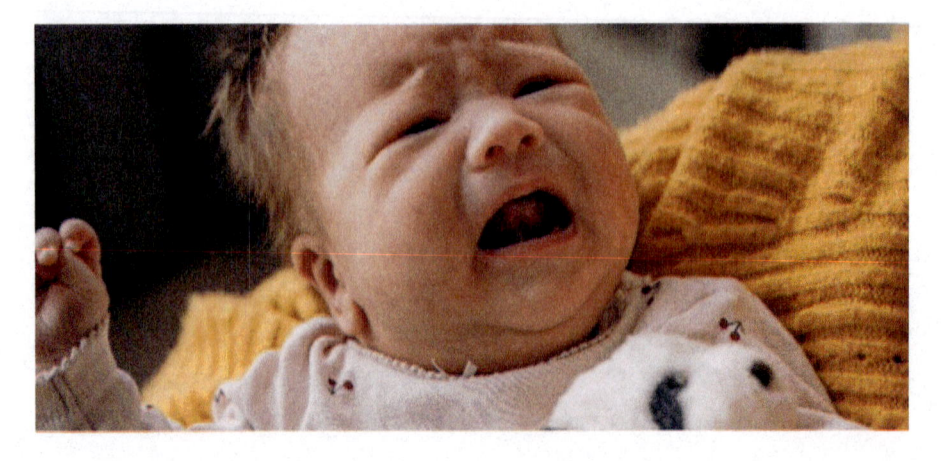

Is my baby allergic to cow's milk?
While Cow's Milk Protein Allergy (CMA) is possible, it is not the most likely cause of crying and fussing.

Numerous new mothers, in their desperation, read that their baby might be allergic to cow's milk in their adult diet passing into breastmilk, or present in baby's infant formula. Please talk to your GP about symptoms before drastic changes to your diet or buying specialist formula. True exclusion diets are difficult to achieve and might have quite an impact on the mother's own health and wellbeing. Specialist infant formulas are not recommended unless prescribed by a doctor.

The NHS website tells us that CMA occurs in anything up to 7% of children under one, but most grow out of it by age five. It is more likely to be experienced when starting a child on solid foods, and is uncommon for breastfed babies.

CMA results in symptoms of an allergic reaction including:

- Skin reactions – such as a red itchy rash or swelling of the lips, face and around the eyes.
- Digestive problems – such as stomach ache, vomiting, colic, diarrhoea or constipation.
- Hay fever-like symptoms – such as a runny or blocked nose.
- Eczema that does not improve with treatment.

Sometimes these reactions occur at the time of exposure, and in other cases they take hours or even days to show.

Occasionally CMA can cause severe allergic symptoms that come on suddenly, such as swelling in the mouth or throat, wheezing, cough, shortness of breath, and difficult, noisy breathing. A severe allergic reaction, or anaphylaxis, is a medical emergency – call 999 or go immediately to your local hospital A&E department.

Has my baby got lactose intolerance?

Lactose intolerance is theoretically possible, but extremely unlikely. Lactose is a normal sugar found in milk and nearly all babies cope well with lactose.

A small number of people have a congenital issue that prevents them from properly digesting lactose, while others develop lactose intolerance later in life. If you developed lactose intolerance as an adult, this is **not** the genetic form that you could pass to your baby.

Symptoms of intolerance include diarrhoea, vomiting, stomach rumblings and pains, and wind. However, most babies will have these symptoms a lot of the time, especially temporarily after a tummy bug! Posseting is normal, babies have runny poos, and most babies have wind!

A truly lactose intolerant baby will be very unwell and not thriving. The baby may struggle constantly with explosive, watery, foamy, frequent green poos and will need to be properly assessed by medical professionals.

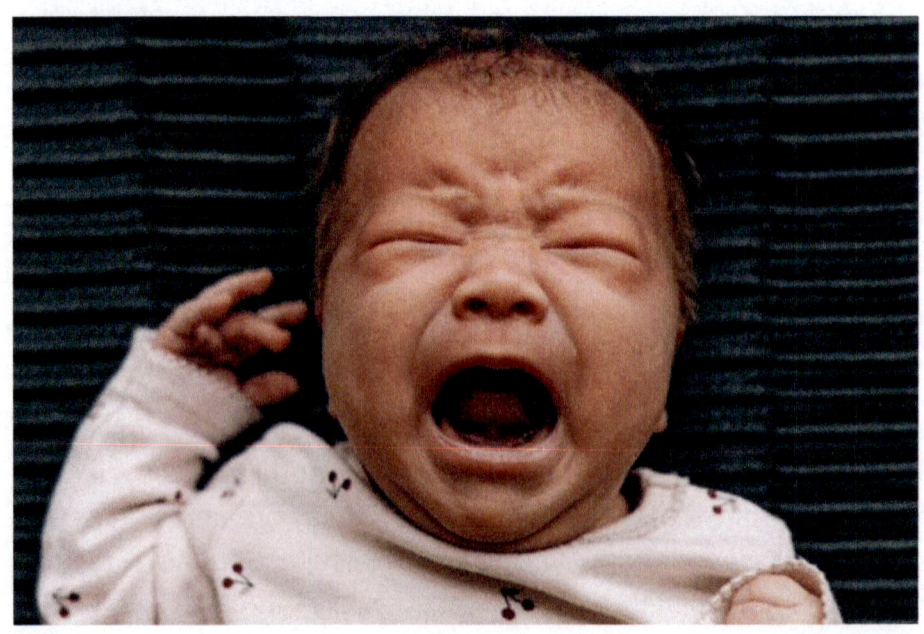

Has my baby got colic?
Nobody knows what colic is.

You might be told that your baby has colic. As Dr Jack Newman says, "*Colic is one of the mysteries of nature. Nobody knows what it really is, but everyone has an opinion.*" Very often a baby starts to have crying periods about two to three weeks after birth. This mainly happens in the evening. The baby may be inconsolable, although walking, rocking or driving with the baby might help for a while. Otherwise, the baby is well and putting on weight.

Colic stops around three months old, sometimes a little later. We do not know why it happens. There is no proven treatment for colic. Many medications and behavioural strategies have been tried, and none proven to work. That includes over the counter medications such as "Infacol" and "Colief". We will all know someone whose baby was "cured," and most treatments work sometimes or for a short time. To be honest, the baby was likely to grow out of it anyhow.

Has my baby got reflux?
Reflux is when a baby brings up milk, and it is totally normal in newborn babies.

Almost all babies vomit or posset up milk at some point in the day, and some do it many times a day. It is to do with the immaturity of their digestive system. It will probably happen more if your baby takes a lot of milk or milk quickly from a rapidly squirting breast or fast bottle feed. It may happen if you move your baby around vigorously after a feed, such as during a nappy change. Sometimes it happens between feeds when your baby is lying down in a cot or pram. If your baby is putting on weight, thriving and doing an appropriate number of wet and dirty nappies, this is just a laundry issue.

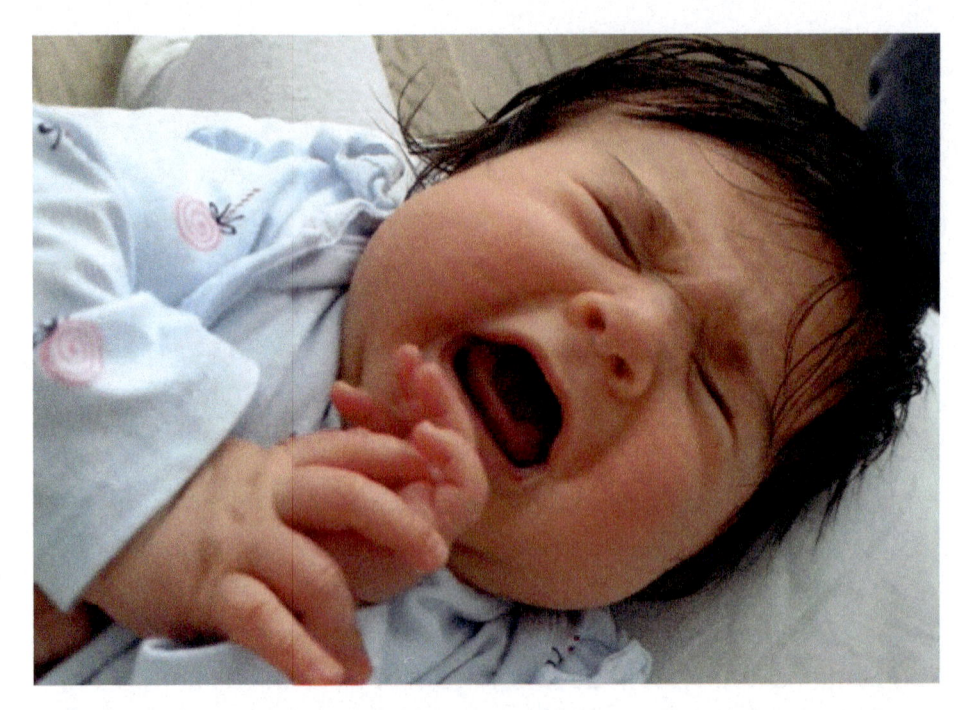

What can you do about it? See a lactation professional if fast flow is an issue, perhaps feeding your baby against gravity in a laid-back position might help. Feed your baby on one side until they are not swallowing milk, then switch sides. Pace your bottle feeds using responsive feeding techniques. Use a slower flow teat perhaps. Your baby will grow out of this, as in most things.

"But my baby isn't thriving. Feeding is a nightmare, and my baby is dropping centiles on the growth charts".

Now this is a different thing. If your baby is not doing well with feeding we need to know why, and I would recommend seeing your GP and also some feeding professionals to rule out medical and mechanical issues.

These babies often seem to be writhing in pain and many parents are offered antacids (that seem to cause constipation) or even medications that reduce stomach acid. However, this is also controversial and there is limited evidence that they work.

As babies are regularly fed milk, their vomit is not particularly acidic. Investigations rarely show inflammation or damage to the oesophagus. True Gastro-oesophageal Reflux Disease (GORD), that we experience as adults, is unlikely. The medications prescribed for reflux can have quite significant side effects, not least uncomfortable spikes in stomach acid production as we take our babies off the drugs. They should be used with caution. The "Discontented Baby Book" by Dr Pamela Douglas offers a useful perspective on this, and how we look after parents' mental health while it is going on.

If a newborn baby has started projectile vomiting every feed, not just the occasional one, they need to be urgently reviewed by a medical professional in case something is wrong such as pyloric stenosis (which is very rare).

Is my baby tense after the pregnancy or birth?

Some babies struggle to feed or stay comfortable if they have a tight neck or jaw, or muscle tension elsewhere in their body. These babies often have a crunchy suck that causes nipple pain when breastfeeding. Sometimes their body feels tense or rigid. An experienced paediatric osteopath might be able to help.

I have worked with many babies that are uncomfortable when feeding, will only feed well on one side, want to feed but seem to be in pain, or start the feed before arching back and refusing to relatch. When we look closely at these babies, we can see that they have a clear preference for turning their head to one side. We will often notice marked facial or head asymmetry. These babies are unable to keep their body relaxed and straight. When lying down they will immediately bow into a banana shape or look off to one side.

Why does this happen? Often these babies were breech for a long time, or had their head back as if star gazing. Most had a protracted labour and birth due to being in a difficult position. Many of them were ultimately brought out with instruments or by caesarean.

When a baby is very tight on one side of their neck it is called torticollis. We can help these babies with osteopathic treatments, and by gently encouraging them to look the other way by waving toys on the less favoured side, varying their direction in the cot or on their changing mat, and slowly moving objects from the preferred to difficult side to encourage them to follow with their head.

What can I do if my baby suddenly won't feed?

If your baby is well, try not to panic, this is usually a short-term thing. You might need to find creative ways to smuggle milk in, often when your baby is nearly asleep.

Some breastfed babies suddenly stop wanting to feed. This can be both frightening and upsetting for their mother, who is obviously desperate that they feed and stay healthy. You might hear this referred to as a "nursing strike."

Why do babies do this?

With babies, it is often hard to tell. Sometimes a nursing strike is associated with a baby being unwell, for example having an ear infection or sore throat. Sometimes it is a new tooth coming through. Perhaps the baby is just getting older and would prefer to catch up on all their breastfeeding at night while they play all day. Sometimes it might be linked with a change in routine, such as going to nursery. Perhaps mother is pregnant or ovulating, or on new medication, any of which may temporarily reduce supply. Maybe she shouted at her baby when they bit her while feeding. It can be linked to very fast let down or oversupply that overwhelms the little one. Sometimes we just don't know.

Please try to remember that your baby loves you and is not rejecting you. Nursing strikes are not a sign that it is time to wean onto solids or switch to formula milk.

How long will it last?

The good news is that most babies on nursing strike seem to stay perfectly well, continue to do wet nappies, and start breastfeeding again when they feel like it! Many of them are taking milk at night. Sometimes the strike is a few hours, but it can last a few days.

What can I do about it?

If your baby is unwell, the first thing is to get some medical attention or pain relief for your little one. Tiny babies (under three months) who are clearly unwell need to be seen by a medical practitioner as a matter of urgency. An older baby with obvious signs of teething could be medicated at home.

Provided all is well, try not to panic. The thing is not to force your baby to drink, but to tempt them whenever possible. We want nice things to happen at the breast. Skin to skin contact and lots of cuddles might help.

In the meantime, please protect your milk supply by expressing (ideally around usual feed times).

Most mothers have some success when their little one is sleepy, such as before and after naps.

If the strike continues for many days, as always you will need to keep your baby fed. Please reach out for support.

Tips for nursing strikes

- Feed while on the move, with your baby in a sling, or hold them and also bounce on an exercise ball.

- Feed when your baby is super sleepy, or in their sleep.

- Have a bath together and offer a breastfeed.

- Lie in bed topless, playing with your baby, with the breast on offer if they want it but with no pressure.

- Express a tiny bit before offering the breast, so that a let down is already happening as your baby starts to feed.

- Vary nursing positions. Baby might have grown out of something that you are doing. Babies with colds or ear ache might temporarily find some positions less comfortable.

- Consider sleeping together safely so that your baby has access to the breast.

- You might offer your baby milk from a bottle in a paced way.

My baby bites me! What can I do?

This is usually a temporary problem. Give your little one something more appropriate to chew on. If they bite, calmly stop the feed.

I receive a lot of calls from mothers whose older babies have bitten them. They are, understandably, upset and dreading the next feed. Some are concerned that they might have to stop breastfeeding before they feel ready, others ring me to ask how to stop immediately.

It might help you to know that breastfeeding is designed to continue even when your baby has a mouthful of teeth, and that when they are actively nursing they can't bite because their tongue has to cover their bottom gum. However, biting is still a common phase which generally does not last more than a couple of weeks.

Biting happens most as the first tooth appears. Some babies are really interested in how to use this new thing in their mouth, and they experiment with a bit of a bite. If their mother says, "no biting" and calmly stops nursing, or more likely screams out in pain, they tend to stop doing it pretty quickly.

Screaming at your baby is not the recommended strategy because some babies panic and then refuse to nurse. But, frankly, it is a pretty normal reaction.

If your baby bites, here are some things that might help you.

Remember that it is usually temporary!

It is common for babies to bite for a while, the phase usually lasts from a few days to a couple of weeks.

Biting is soothing for your little one's sore mouth – so give them something more appropriate to chew on.

There are many teething toys on the market that you could offer your little one outside of feed times, or if they look like they are about to bite, or if they have bitten and you stopped the feed.

If biting becomes a habit, think about when it happens during the feed and be ready to detach your baby.

Biting at the end of a feeding session – be alert and ready to intervene.

Many mothers find that their babies feed fine while the milk is fast flowing and their babies are busy drinking, but that they bite when they get a bit bored or frustrated towards the end of the feed. Some women take care to watch their babies at this point, and have a finger ready to detach their baby when the sucking slows down. It might be a good to swap breasts a little sooner, to work with the fastest flow available. Praise your baby when they don't bite, and offer a teething toy to chew on when they start to look like they might!

Biting at the start of a feed.

You might consider offering a teething toy before a feed, and praising your little one when they then latch on well and don't bite. If they are distracted and fidgeting, probably best avoid feeding and do something else to calm them first. Tips for feeding a distracted but hungry baby include actively rocking, bouncing on a birth ball, walking (if they aren't too heavy) or going to a quiet room for a lie down with them.

If your baby bites for attention.

Sometimes a baby will bite just to get mummy to focus! They might even think it is a brilliant game. The only real advice here is to be on guard. If your baby bites, end the feed, and either say nothing or calmly try something like "ok, you don't want to feed now?" Stopping the feed is the best way to show that feeding and biting do not go together. If you have stopped the feed, try offering a teething toy, or restart after a few minutes if your baby is keen to continue drinking.

If your baby is older and receptive, you might deliberately distract your baby with something else while they are nursing.

It is not a great idea to scream at your biting baby on purpose.

Try not to shout as a training method. Babies either think it is a game, or get too fearful to nurse. You might yelp accidentally, and don't beat yourself up over this.

If your baby won't let go!

The ideal way to break the seal is to wiggle your finger into your baby's mouth and between their gums. If that does not work, bring baby in much closer to you so that they actively must let go. Obviously don't put your baby in any distress or danger! Some mothers gently pinch their baby's nose for a second to get the same effect.

If you want to stop breastfeeding.

I understand. You might find that this is absolutely the last straw. Or perhaps you were thinking about stopping anyway and this feels like a sensible time to do it. You might consider handing your baby to someone else for a bit, and have a drink or a walk, and have a think about it. Most women prefer to stop breastfeeding on a high point rather than a low one. If you plan to stop, it is kindest to your breasts to wind down gradually.

Skip a feed a day, wait a few days, skip another one. Or do half feeds and top up with formula. If you want to go "cold turkey" that is fine but be prepared to express to comfort if you feel full and engorged, so that you don't risk blocked ducts or mastitis.

Some babies don't bite as much when things are calm at night, or at other times you might notice, so you may consider retaining these feeds for as long as you both feel comfortable.

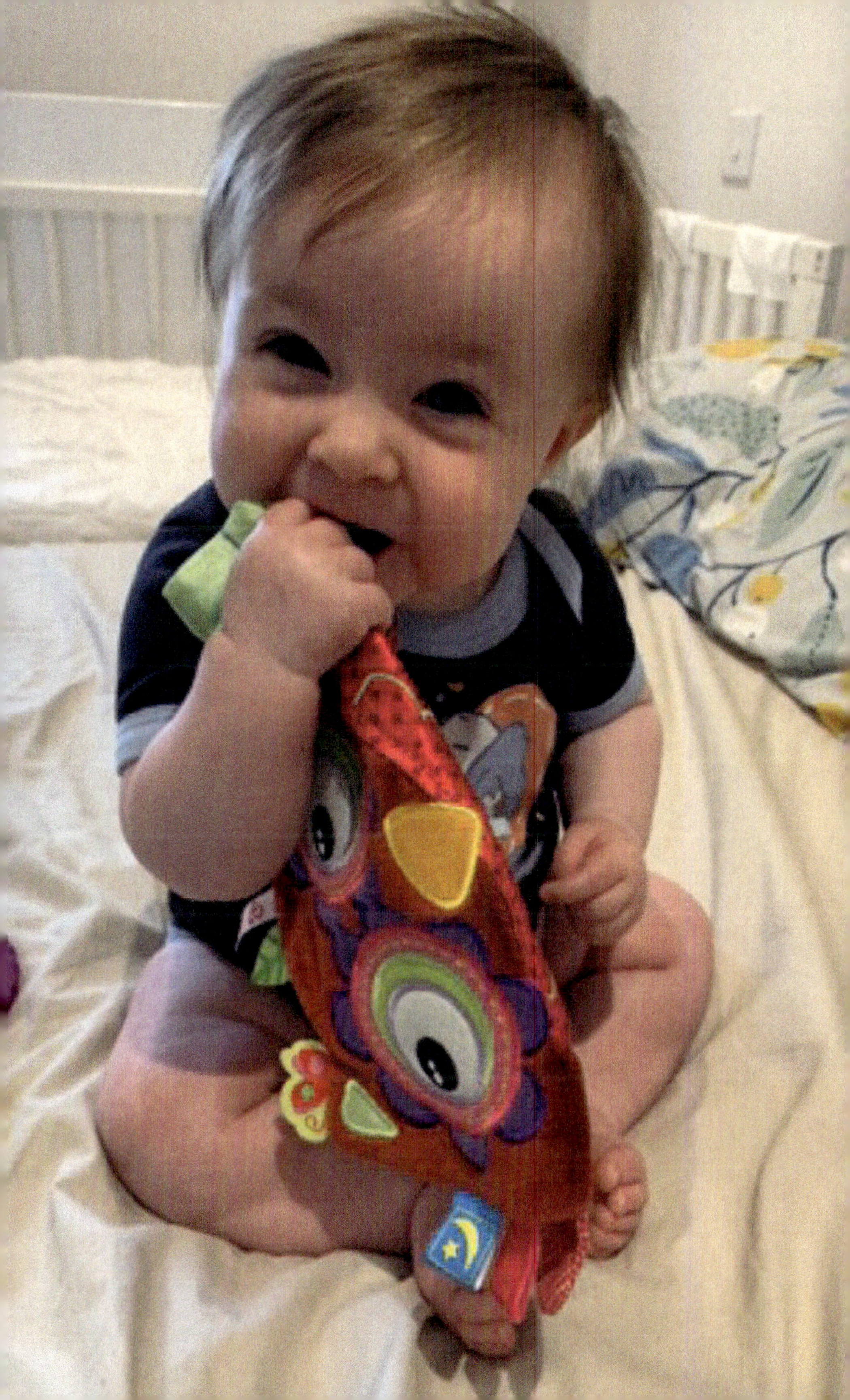

Feeding twins or triplets

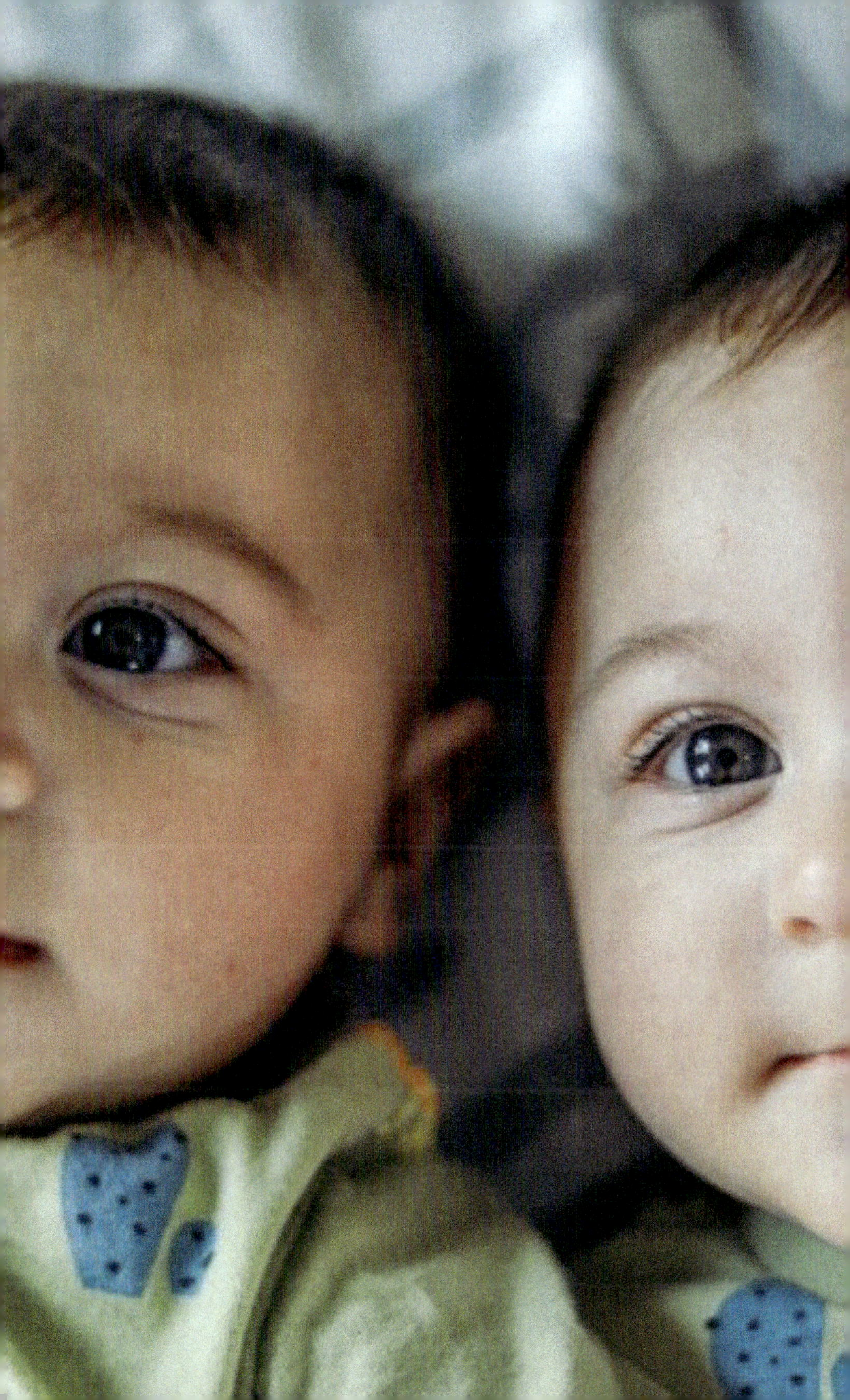

Breastmilk is a wonderful gift and natural medicine for your babies. If you would like to breastfeed, please take advantage of all possible support to enable you to give your babies as much breastmilk as possible. This section discusses some of the possibilities and challenges along the way, alongside practical advice on how to manage.

Congratulations on your babies. This is likely to be one of the most wonderful and yet confusing and exhausting periods of your life. We know that breastmilk provides your babies with the best possible nutrition and builds their immunity, especially when they are tiny and vulnerable.

However, twins and triplets often come a little early, and tend to be smaller, sleepier, prone to jaundice, and trickier to feed. It is possible that they may need time in Neonatal Intensive Care (NICU) or transitional support in hospital until you are able to take them home. In situations that feel more medicalised than you might like, providing breastmilk can be a lovely way for you to feel more involved in your babies' care.

Is it possible to exclusively breastfeed more than one baby?
Yes!

When a mother's body is working well, her breast tissue and hormones help her to produce sufficient milk to fully meet the demands of her little ones. As the babies take milk at the breast, it gets replaced. This means that is it often possible to exclusively breastfeed twins or triplets, especially the closer they are to full term when they are born.

Some women find the process immensely rewarding, especially when other parts of their babies' birth and early lives may have been outside of their control.

Realistically, many mothers who want to exclusively breastfeed find that it can take considerable time and effort to get established and requires a lot of skilled and caring support. Once they have got two or more babies breastfeeding, it saves a lot of time, effort and expense over formula feeding.

What are the unique challenges around feeding multiple babies?

Any number of the factors below can interact to make things more difficult for a new mother of multiple babies.

Many women have unrealistic expectations of themselves and their babies, and this is intended to inform and support your decision making rather than to make you feel like the odds are stacked against you.

Factors that might impact milk supply.

Pre-term birth.

A woman's breasts develop the glandular tissue and blood supply they need throughout pregnancy. If her babies are very early, the breasts will not have done all the preparation and growing they need to maximise milk supply.

To put this in context, about 40% of twins and nearly all triplets are born prematurely (before 37 weeks) or unwell and may spend time in the neonatal unit. Babies born before 32 weeks or who are under 2 kg at birth will be cared for in NICU and will need extra special care with feeding.

Medical complications.

Multiple pregnancies can be associated with other medical conditions, including pre-eclampsia and gestational diabetes. Sometime a mother will be feeling too unwell to breastfeed, and the medication required can also impact on her milk supply. Women being treated for severe pre-eclampsia often report feeling "out of it" for the early days of their babies' lives.

Blood loss at the birth.

Significant blood loss during the birth can have a big impact on the mother. She can feel very faint and tired. She may not be able to fully interact with her little ones to feed, or express milk, for hours or days. This can result in reduced milk supply. In many cases there is a temporary period when supply is lower than it should be, and supplemental donor milk or formula will be needed. Sometimes a severe bleed can result in the pituitary gland in her brain being unable to make the prolactin required to get milk production off to a good start. On occasions her body will not be able to produce sufficient milk in the long term.

Recovering from caesarean section.

When a mother is recovering from abdominal surgery it can be hard for her to move around and attend to her babies. This can result in a delay in milk coming in, making it more likely that the babies will need supplemental feeds in the meantime.

Exhaustion.

Caring for multiple babies around the clock is incredibly demanding. When a mother has so little time to rest, recover and look after herself, milk supply is sometimes compromised. Mothers with wider work or family commitments, or who have been separated from their babies, may find it especially difficult to feed and/or express sufficiently to maintain a full milk supply for more than one baby.

Factors affecting the babies' ability to breastfeed.

Ability to suck and swallow.

Babies learn to suck in the womb, and this becomes more effective after about 34 weeks' gestation. If the babies are born before this, their suck reflexes are not yet sufficiently mature to breast or bottle feed, and they might require feeds through a tube into their nose and down into their tummy. As they grow, they will gradually be introduced to the breast or bottle and they will learn to coordinate their suck, swallow and breathe pattern.

Tiny mouths.

Even if they can suck, tiny babies tend to have little mouths, and sometimes they simply cannot fit around their mother's nipple. In these cases, babies require expressed milk or formula from a narrow teat until they can latch on directly. Mothers with flatter nipples might find that a small nipple shield can help a little baby to maintain their attachment to the breast.

Sleepy babies.

Premature babies can often be excessively sleepy and use up their energy reserves quickly. They are also more likely to get jaundiced, which can make the situation even more challenging as they are both sleepy and sometimes in need of phototherapy. In many cases, tiny sleepy babies need help in getting their feeds efficiently and effectively.

Need for medical attention.

Babies may need some extra medical support, especially if they are very small. While many hospitals promote skin to skin care when possible, the need for treatment and monitoring can make it harder to move your babies around and feed them directly.

How can I get off to the best possible start with breastfeeding?

You can prepare to breastfeed your little ones before they are born. It is very sensible to find out about local sources of feeding support.

In terms of kit, you might consider where to hire or purchase a double breast pump, and gather some narrow teat feeding bottles and a steriliser. Many parents suggest obtaining a twin feeding pillow and at least one bouncy chair (for one baby while you deal with the other). Your local twins club can be an excellent source of peer support and second hand items.

After about 36 weeks of pregnancy, you could try hand expressing colostrum. Your little ones may benefit from the colostrum soon after birth, and it is a useful skill to practice in advance. Midwives can provide the collection syringes and instructions on how to do this. Squeeze behind your nipple and collect any drops of colostrum into a feeding syringe.

You can pop them into a food container and store them in the freezer. Ask your midwives about when to bring this milk into hospital, and where it can be stored if, for example, you have a planned caesarean section. Most postnatal wards have a special milk fridge for this purpose. Your milk will need to be properly labelled with sticky bar codes (usually kept in the back of your notes), and the date and time added.

New parents of multiples benefit from hands on support to cope with the physical and logistical as well as the emotional transition to new parenthood. Mothers of multiple babies commonly have a caesarean section, which is major surgery, and it takes significant time to recover. It is a great idea to prepare family and friends, or source paid supporters such as postnatal doulas or night nannies to help in the early weeks.

Immediately after the birth.

It is a great idea to hold your babies skin to skin after they are born and for as long as you can in the early days and weeks. This helps them to feel relaxed and happy, regulates their temperature, be colonised by your bacteria, and be in the perfect place to feed. When the babies are tummy to tummy with you, with their mouths next to the nipples, they are primed to use their natural instincts to start to breastfeed.

Ask your midwives for help in getting your babies feeding within an hour to two of birth. If skin to skin contact is not possible, please do not worry, there will be time for that when you are all feeling better. Your babies will enjoy skin to skin with another special person if they are well, but you are not yet up to it.

The first few hours and days.

Your babies will need regular feeds, at least every three hours from the start of one feed to the start of the next. It is very likely that you will work with your care givers on a feeding plan, especially if your babies are small or clinically vulnerable. Keep clear notes of which baby has done what, in terms of feeds and nappies. There are apps available to help with this. Make sure that everyone involved in your babies' care also adds to the notes.

If your babies cannot breastfeed.

If your babies cannot breastfeed effectively, you have two main priorities. The first is to feed your little ones somehow, and the second is to build and maintain your milk supply until they can breastfeed.

If you have premature babies in the NICU the staff will talk to you about the most appropriate way to feed them. If your babies are not able to suck and swallow, or be moved from their incubator, they will be fed through a tube that goes through their nose and into their tummy. Most hospitals promote "kangaroo care" when possible. This means that the babies are cuddled against their parent's skin to skin for as much time as possible. We know that this decreases stress and promotes healing in the newborn, and it is lovely for the carers too.

The ideal food for all babies is breastmilk. If you can express colostrum, you will be invited to do this by hand until your milk comes in. You will be encouraged to express as regularly as possible, from both sides, ideally every three hours including at least once overnight. The nurses can help you and provide the feeding syringes to collect the milk. Once you have more than 5 ml per side, you can use an expressing machine. Ask the nurses for a hospital pump and the kit required to express and store your breastmilk.

Short, frequent expressing is more effective than one or two longer sessions. If one or more babies are also being breastfed, it makes sense to express after the breastfeed. Women who have been discharged home might benefit from hiring a hospital grade pump from the manufacturer or a local pharmacy.

Breastmilk and breastfeeding are known to be an effective form of pain relief for babies, so might be encouraged during medical procedures or immediately after them.

If a mother is not able to breastfeed, or fully breastfeed her babies in NICU, it might be possible for the little ones to receive carefully screened milk from a donor. This is the next best thing to mother's milk, and helps to protect babies from complications including necrotising enterocolitis (inflamed intestines).

In cases where no breastmilk is available, the hospital will provide infant formula suitable for your babies depending on their gestational age.

Transitioning to breastfeeding.

As your babies get older and more able to take milk at the breast, you might be able to breastfeed at least some of the feeds, or a part of each feed. While the babies make this transition, it might be appropriate for them to receive additional milk through a feeding tube into their tummy, finger feeding, or supplemental nursing system taped to the nipple while they are at the breast. In this way they learn to associate breastfeeding with a comfortable feeling of being fed. On other occasions, it might be sensible to offer the breast first, and then provide a top up of expressed milk, donor milk or infant formula from a bottle (see below). Infant feeding specialists and NICU nurses will be able to help and advise in this situation.

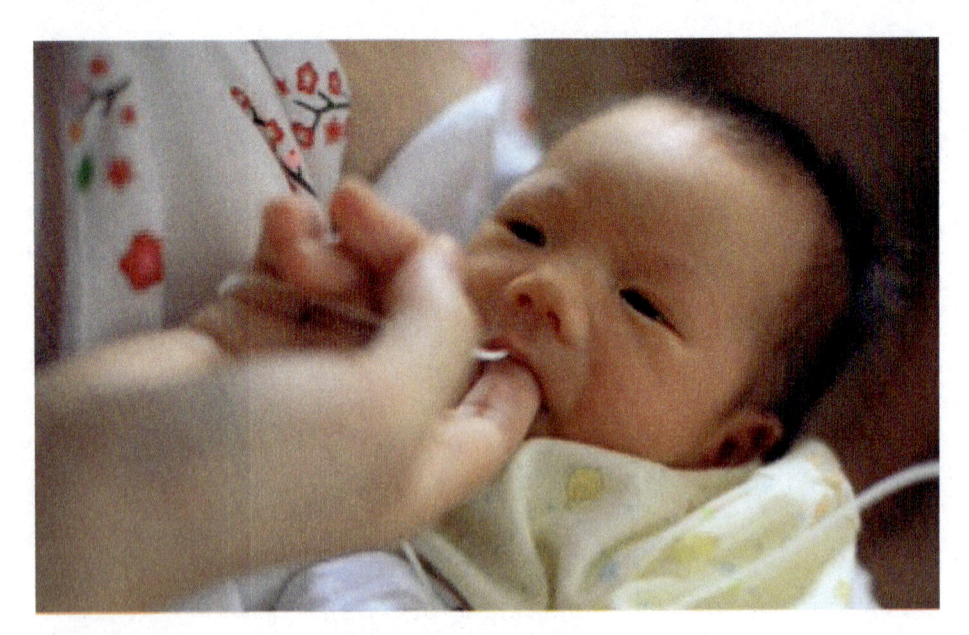

Topping up your breastfeeding babies.

You might find yourself in a situation where your babies are receiving some breastmilk direct from the nipple but still need extra top ups after some or every feed. You might be asked to breastfeed, top up your little ones with expressed milk, and then express in advance of the next feed.

Most new parents do not find a cycle of breastfeeding, topping up and expressing sustainable for long. There could be a balance to be found between time at the breast and time spent expressing. It might be sensible to breastfeed while you can see that your babies are actively swallowing milk, and then take them off to be topped up by a partner while you focus on expressing. Every case is different. The babies and mother will benefit from time skin to skin while practicing feeding. However, the mother's needs for rest and recovery should not be overlooked, especially after a surgical birth.

Another approach to top ups is to offer the babies one or two whole feeds of expressed or formula milk at certain times, and breastfeed responsively in between.

Consistency helps here, as the breasts soon learn what is required of them. For example, it might be possible to breastfeed the babies through the mornings and always offer a bottle say mid-afternoon and/or in the evening before bed. This may have the advantage of giving the mother a proper break, and allowing her to continue to breastfeed at other times.

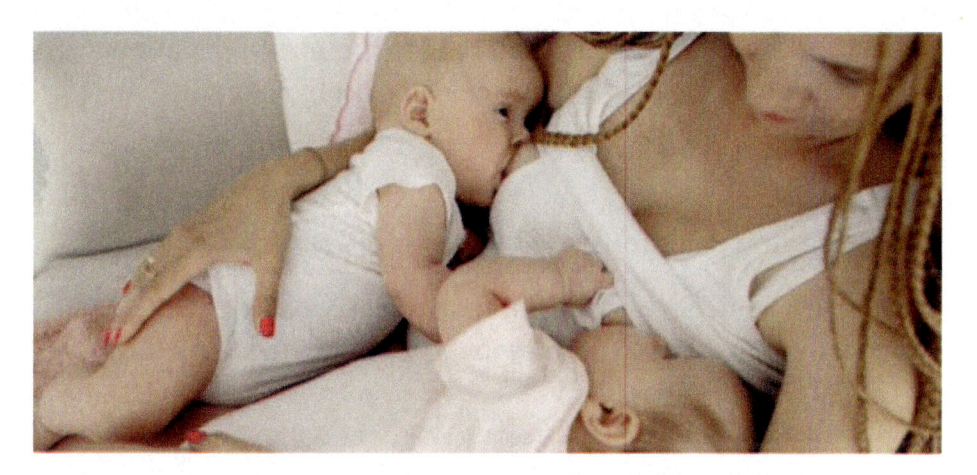

My babies are breastfeeding! Now what?

It is often the case that one breast naturally produces more than the other, or one baby is better at feeding than the other(s). Generally, it helps to swap around each feed so that all babies stimulate and receive milk from both breasts.

Do I feed my babies together?

When you are all learning, it can make sense to feed the babies individually. However, this takes considerable time, and you might feel like you do nothing but feed! When both twin babies are effective feeders, latching them on together saves time. It takes practice though, and you either need someone to pass you the babies once you are comfortable or a system where you can sit down and still reach them. Feeding pillows and bouncy chairs can be very helpful tools. Firm, flat feeding pillows are better than squishy rounded ones.

What about triplets?

There are no hard and fast rules. Some mothers of triplets feed two together and then the next individually, and rotate the pattern. If the woman has help, some choose to feed two together while the third is offered a bottle of expressed milk or formula. Others single feed one baby and express afterwards for two. It might be sustainable to single feed one baby on rotation and formula feed the other two. Anything that works! Combination feeding breastmilk and formula still provides lots of the benefits of breastfeeding.

Which feeding positions are suitable for two babies?

Various positions allow a mother to feed two babies at once. The most popular include the rugby/football hold where both babies are held underarm, or the "koala" hold where both babies are angled down the mother's body next to each other. While a mother recovers from a caesarean, she might prefer the underarm hold as it keeps her babies off her bump and away from her scar.

Certain positions, including the rugby/football hold, are much easier with a good-sized feeding pillow so that one baby is held in the right place while the mother manoeuvres the other one. It helps to have another adult available to pass the babies to mum.

Once a mother has recovered from the birth, which might take several weeks, she may benefit from having one or two newborn bouncy chairs for the babies so that she can keep them close to her while getting into feeding positions. As your babies get older a range of positions become possible, and many women get creative. I worked with a mum who fed one baby in a sling and the other at the same time in koala hold.

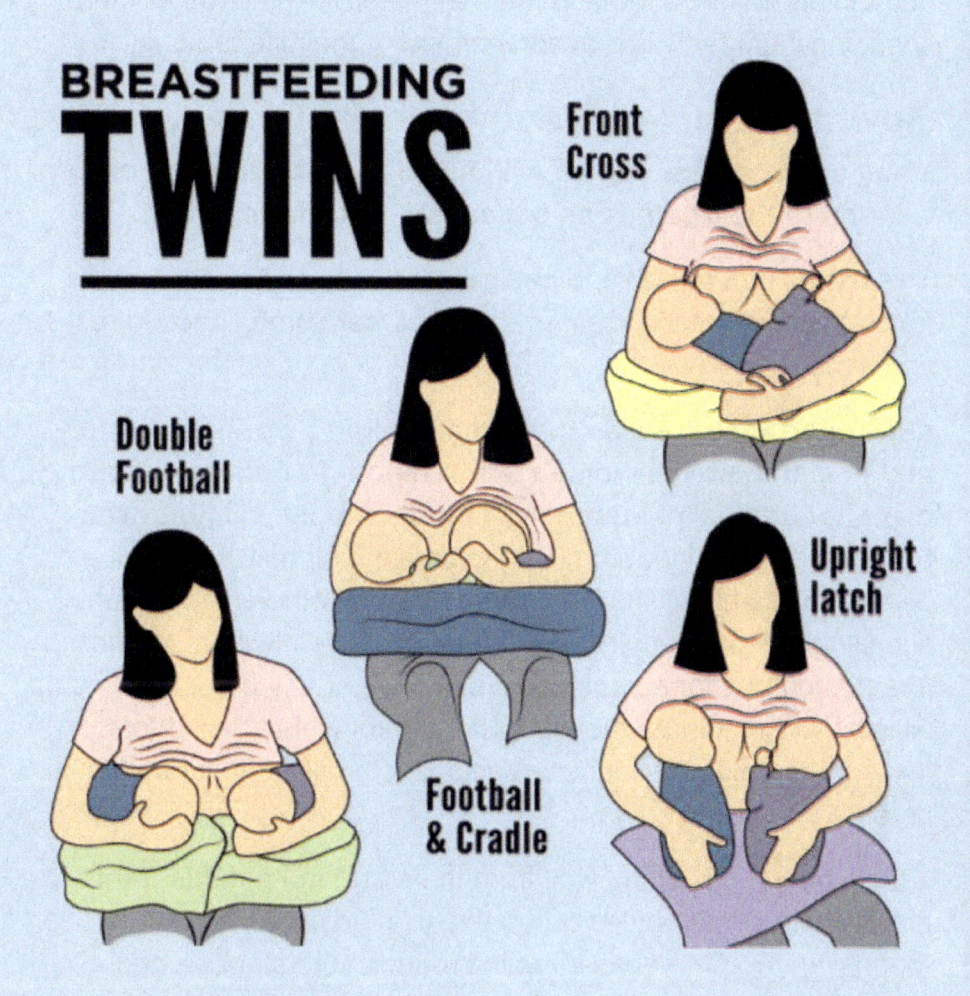

BREASTFEEDING TWINS

Front Cross

Double Football

Football & Cradle

Upright latch

Do I wake my babies to feed them?
It might be necessary to wake your tiny babies on a schedule especially if they are excessively sleepy.

Later you could be more baby led, but many parents of twins and triplets insist on feeding them at the same time rather than spending what feels like every moment of their life feeding!

How does mixed feeding work?
Many parents of multiples will mixed feed, offering formula top-ups to their babies as necessary or on a schedule.

This allows other people to get involved with feeding the babies, takes pressure off the mother and gives her some space to rest and heal.

It is important to realise that breastfeeding works on supply and demand, so if a woman is not feeding or expressing to meet her babies' needs then her milk supply will go down. Some women find that they can maintain their milk supply at a certain level while also using some formula, others find that their supply tumbles to a point where they consistently have to offer formula top ups for each feed. We know that any breastfeeding at all is beneficial for the mother and the babies, but mixed feeding takes more time and causes additional complication for some parents.

If breastfeeding is going well from the start, and parents want to introduce some bottles of formula, it is advisable to wait for roughly four to six weeks for milk production to stabilise and everyone to get confident with breastfeeding before introducing formula. Many women find that their milk supply is more flexible after a few weeks, and will adjust well to some missed feeds. It is so individual. Some women will need to feed or express a certain number of times a day to keep up supply.

Many parents find that offering a formula bottle in the evening has the least overall impact on milk supply. Giving formula in the early hours is often counterproductive, as a woman's milk supply is highest at this time so it is easier to feed or express, and her breasts might be uncomfortably full if she misses a feed.

If a certain number of feeds in a day are going to be formula from a bottle, it makes sense to alternate the bottles with breastfeeds/expressing so that the breasts do not remain full for excessive time. Leaving the breasts full for extended periods causes an associated reduction in milk supply and discomfort for the mother.

How much milk will my babies need?

This will vary depending on the age and size of your babies. Your midwives or neonatal nurses will help you to work it out.

Obviously, we cannot tell exactly how much milk babies take from the breasts, we can only get an idea from watching and listening to swallows and looking at the output in the nappies. We look out for feeding cues (opening the mouth, trying to suck fingers, trying to suck anything, crying) and respond to our babies.

Many of you would like more certainty around quantities. Please figure this out with your medical team if your babies are preterm, and remember that the amounts will need to be age adjusted. For full term babies, after the first few days, paediatricians would suggest approximately 150 ml milk, per kilo of birth weight, in 24 hours. From one month to six months, most term babies would take 750-900ml per day.

Formula manufacturers use an equation for amount per feed by age and number of feeds, and this is written on the packaging. Again, this is a guide, and we need to be led by our babies to an extent.

How can I cope with feeding my babies?

Parents of multiples will tell you that feeding them is a full-time job for at least two people! Please get as much help as you can.

Keeping these little babies fed, clean and rested is all you can manage, and looking after yourselves can seem impossible.

Ask for as much support as you can find, even putting family and friends on a rota. Most people would love to help, they just might need a little direction on how best to do so.

If you can consider paid support, perhaps look for an experienced postnatal doula or night nanny to help you.

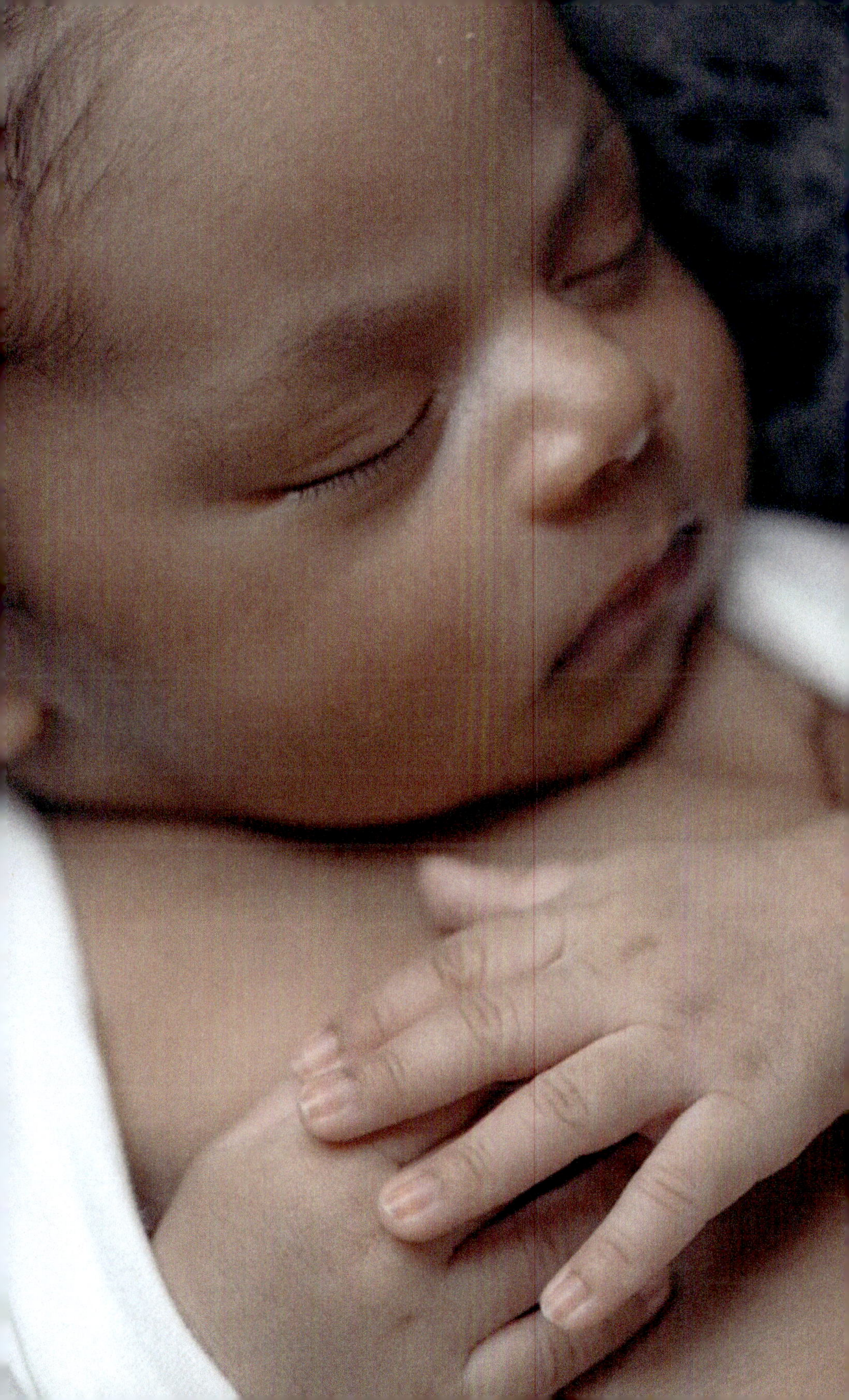

Special situations

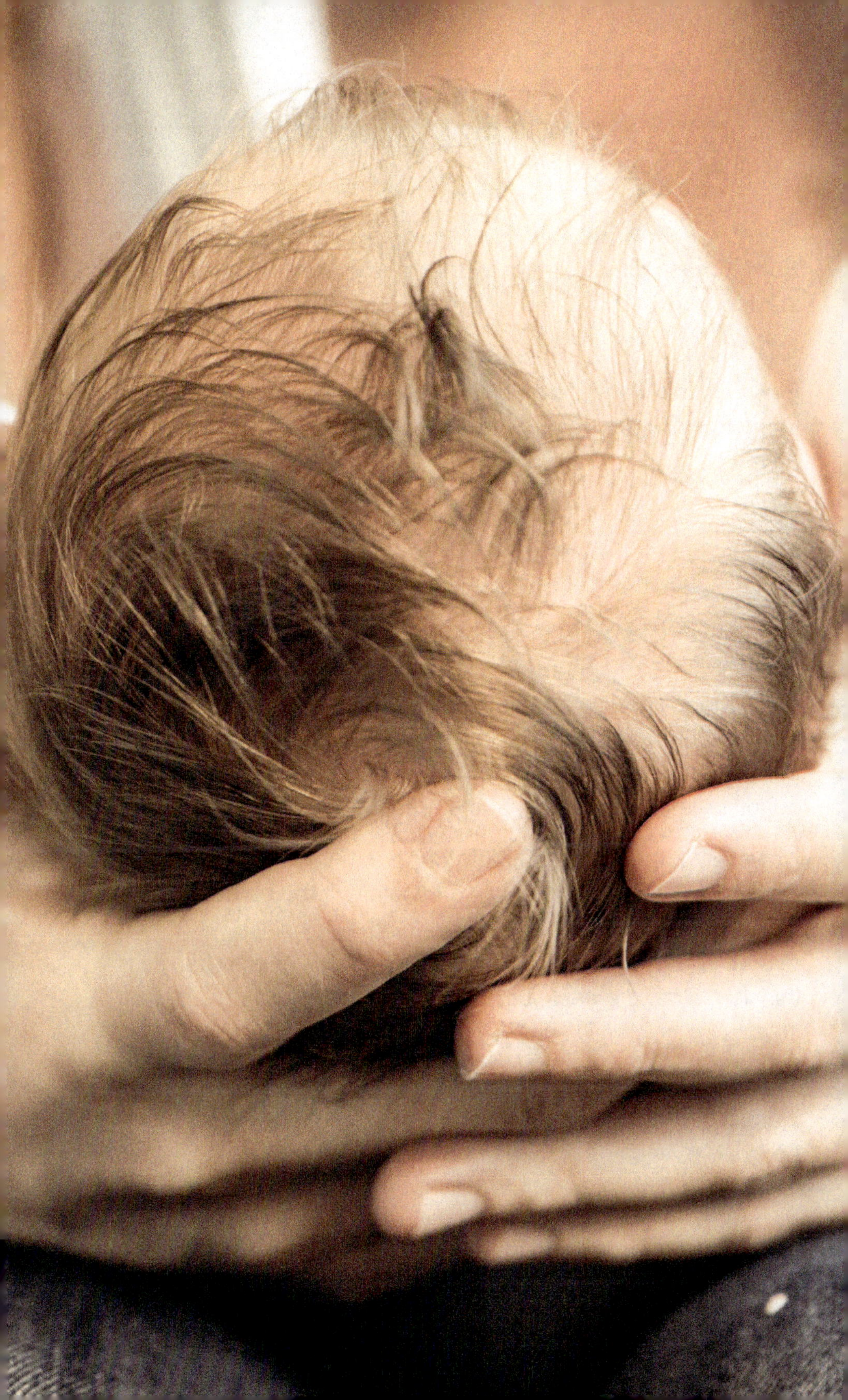

In this section we consider the implications of serious issues with a mother or baby's health.

How can I cope with a difficult start?

Sometimes things do not work out the way we would like. Be kind to yourself and get plenty of support.

"Everyone says that meeting your baby is the most amazing feeling in the world. I was in the operating theatre with some guy trying to stitch me up and I felt like shit. I did not even want her on me."

When most of us picture the birth our baby, we hope that everything goes to plan. We picture relaxing in a haze of blissful oxytocin, gazing at our precious newborn.

The baby books and TV do not prepare us for the moment of meeting our baby for the first time. Sometimes it is one of the most intense and incredible moments of a person's life. After a natural birth with minimal intervention, the body is primed to release a huge boost of love hormones that help to promote bonding, lessen bleeding, and start breastfeeding.

However, it is equally common for parents to feel aloof and detached, "out of it" or unwell, particularly if the birth has been long or frightening or there are ongoing medical issues for any of you. You might spend the first few minutes of your baby's life thinking *"what the heck just happened, and who is this person sewing up my pelvic floor?"*

So, a rocky start is more common that you might think. How could it affect baby's feeding? Well, it is ideal for mother and baby to be close to each other, preferably skin to skin, to calm them both and get breastfeeding started. When this is not possible, we must accept that and do the best we can to bring mum and baby together when they are both well enough.

Hospital staff are completely aware of this but may be overstretched. It might be necessary to ask for help in promoting cuddling and bonding time even with babies receiving special care or neonatal intensive care.

Babies love being with any of their grownups, so if mum is unwell it is ideal for dad, a partner, friend or family member to have contact with the baby including skin to skin contact. This will be a special time for everyone.

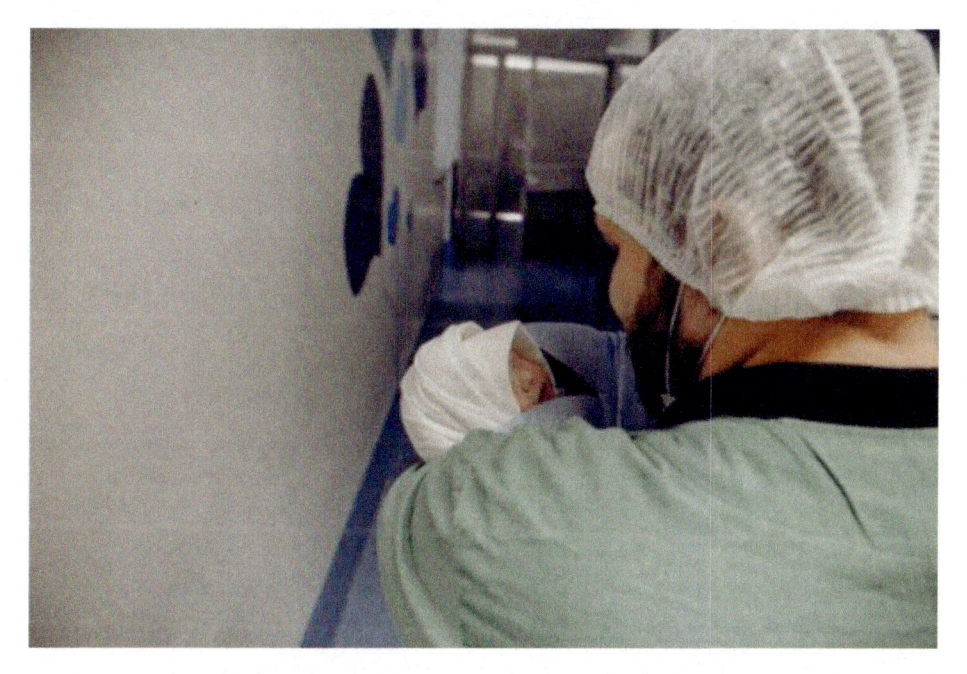

If a mother wishes to breastfeed and is well enough to try it, then she should be offered as much support as possible. If she is not able to breastfeed, it might be appropriate for her to hand express milk for the first three or four days, after which an expressing machine can be used. Expressing milk gives the woman's body the signal that it needs to make more milk, so doing it regularly helps with supply. Aim for every three hours or so, including overnight, but balance this with the mother's need to rest and recuperate too.

What if I am seriously unwell?

If mum is unwell, the medical staff will prioritise her care and her baby will be offered donor milk or formula until she is well enough to breastfeed or express.

There are rare situations in which a mum is unwell after the birth. For example, sometimes after birth a woman will bleed too much. This is called a post partum haemorrhage (PPH). It is an emergency, and lots of people will work to stop the blood flow (for example stimulating the uterus to contract with drugs and compression, locating and sewing up any tears or cuts). The new mother may be given fluids through a canula in her arm or foot, and she may be given extra units of blood. Her partner will be encouraged to talk to her and told the baby close to her. If it becomes necessary to give the woman a general anaesthetic, the partner and baby are asked to leave the room and will be shown into an anteroom to wait until mum is stable and in recovery ward.

When a birth has been traumatic, dad or a birth partner is often left holding the baby for some time, even a few hours. It can be a bonding experience, but obviously also a concerning time waiting for news.

Sometimes the exhaustion, blood loss and medications can make a woman feel completely out of control, unable to process information, unable to look after herself or her baby. Afterwards, women talk about being in a haze, unable to recall the details of what happened. It can be extremely difficult to successfully breastfeed in this situation, and a mother will need a lot of hands on physical and emotional support at this time. There is often a case for bottle feeding the baby if necessary while the mother gets some sleep to begin to heal. A supportive partner or family member can make all the difference here.

It is understandable for the woman to feel that she missed out on special time meeting her baby. It is likely that she will resent her partner for the time that they spent together, even if she does not actually say so. It is possible that she will feel that she, or her body, has somehow failed her. She will require lots of empathetic and careful support.

As with so many things, it can be good to express these feelings. There are various debriefing and support services available through the NHS and privately should they be necessary for either or both parents.

What if my baby is unwell?
There is a slim chance that a baby will be rapidly transferred to the neonatal intensive care or special care unit.

If this is the case, partners are sometimes invited to go with the baby until mum is ready to join them. It might be possible for a partner to take photos or videos of the baby to send or bring back to the mother if they are both well enough.

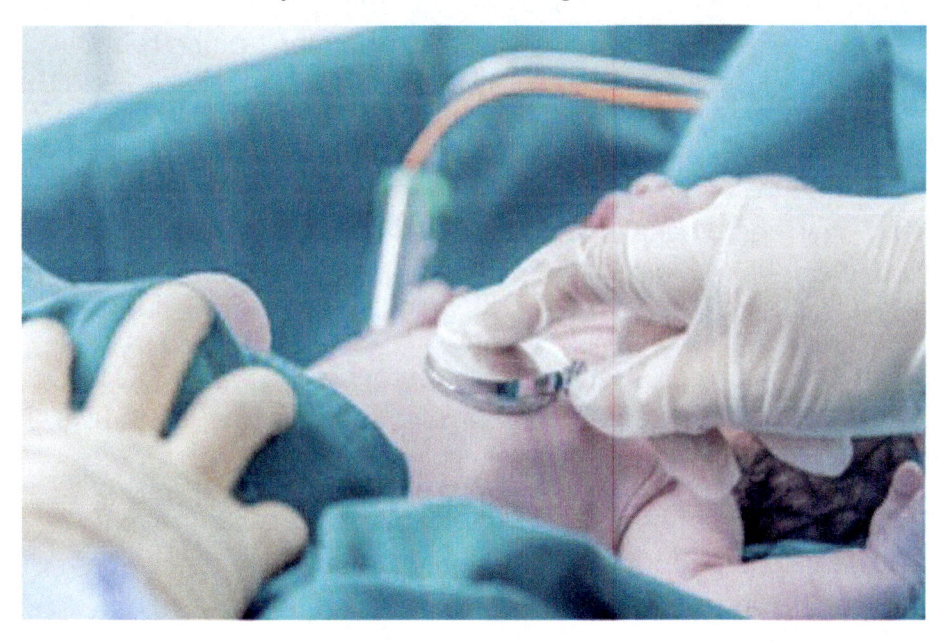

The mother will be supported to go to her baby as soon as she feels able to. The medical staff will support the mother to breastfeed or express milk for her little one if she is well enough. Expressed milk could be fed directly into the baby's mouth through syringe or cup, or through a thin, flexible nasal gastric tube into the baby's tummy. Breastmilk is particularly beneficial when a little one is small or unwell. If the mother is unable to express her milk, the baby might be offered screened milk from a donor if the parents are comfortable with this.

What can I do if I am separated from my baby?

Ask staff for help to see your baby as soon as you are well enough, while making sure that you also receive appropriate care. Partners and friends can be a welcome source of practical and emotional support.

There is an odd set up in many NHS hospitals when a mother is a patient on the postnatal ward but her baby or babies might be in neonatal intensive care or special care. In many hospitals these are separate wards and perhaps in different parts of the hospital.

Mothers want to spend as much time as they can with their newborn(s) but then miss out on the checks, medication and meals that they would normally be receiving on the postnatal ward. This can be challenging for new families and medical staff. It can help to talk to the midwives in charge about when, specifically, it would be good to be in one ward or the other.

If a mother is unwell, it may not be appropriate to walk between the two wards and women may have to ask for someone to take them in a wheelchair.

If a baby or babies need longer term care in hospital, the mother may be discharged home without her little one(s). This can be challenging for many reasons.

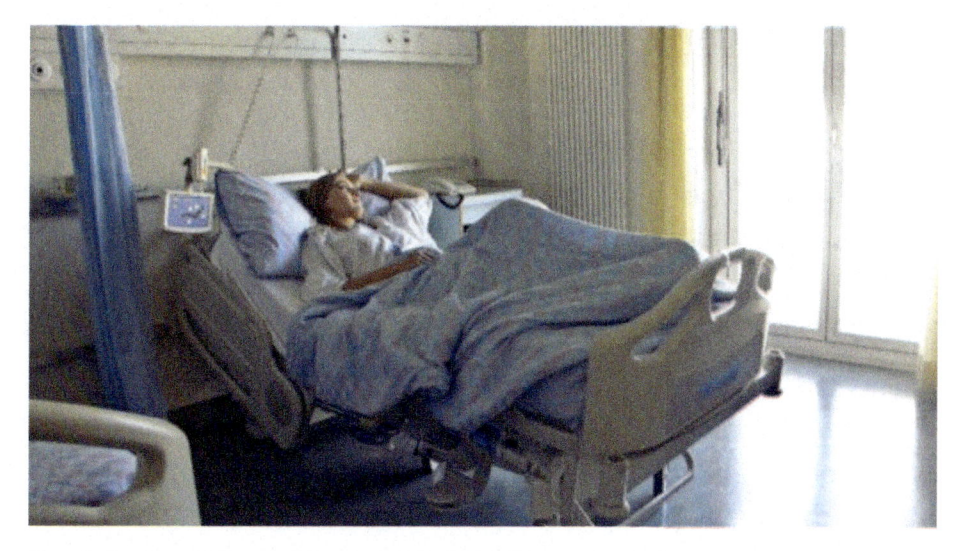

It might be difficult for women who live some distance away, and perhaps cannot visit easily or drive due to a caesarean section. There are many local and national organisations, and hopefully friends and neighbours that can help. The charity Bliss is specifically for parents with premature or sick babies.

A mother will often be encouraged to express breastmilk on a regular basis, ideally 6-8 times a day. Many women find it helpful to hire a hospital grade pump, or invest in a double electric pump. The milk can be stored in the fridge or freezer and taken into hospital by the woman or her supporters.

On occasions, a mother needs to be readmitted to hospital after a time at home. In some cases, within the first two weeks after the birth, a mother will go back to the postnatal ward with her little one. However, there may be times when a mother is a patient in another adult ward and there is not provision for her baby to be with her. Separation of mother and baby is almost never appropriate, and it might be necessary to work quite hard with hospital authorities to make sure that this does not happen unless it is in the best interests of mother and child. This is easier if you have advanced notice, rather than in an emergency.

How do I feed my baby if they are unwell or in NICU?

We know that breastmilk is absolutely the best food for a baby especially one that is compromised and in need of special care. The staff in NICU are trained to help you to feed directly, or to express milk that can be fed to your little one.

The ward will have collection kits for hand expressing and expressing machines, plus milk storage facilities for your named and dated expressed milk. If you wish to breastfeed and your little one is too poorly or tiny to feed at the breast, they may be fed with a small milk feeding syringe, or sometimes a thin flexible tube that goes through one nostril down into their tummy.

Most hospitals promote "kangaroo care" when possible. This means that the baby is cuddled against their parent's skin. We know that this decreases stress and promotes healing in the newborn, and it is lovely for the carers too. Breastmilk and breastfeeding are known to be an effective form of pain relief for babies, so might be encouraged during medical procedures or immediately after them.

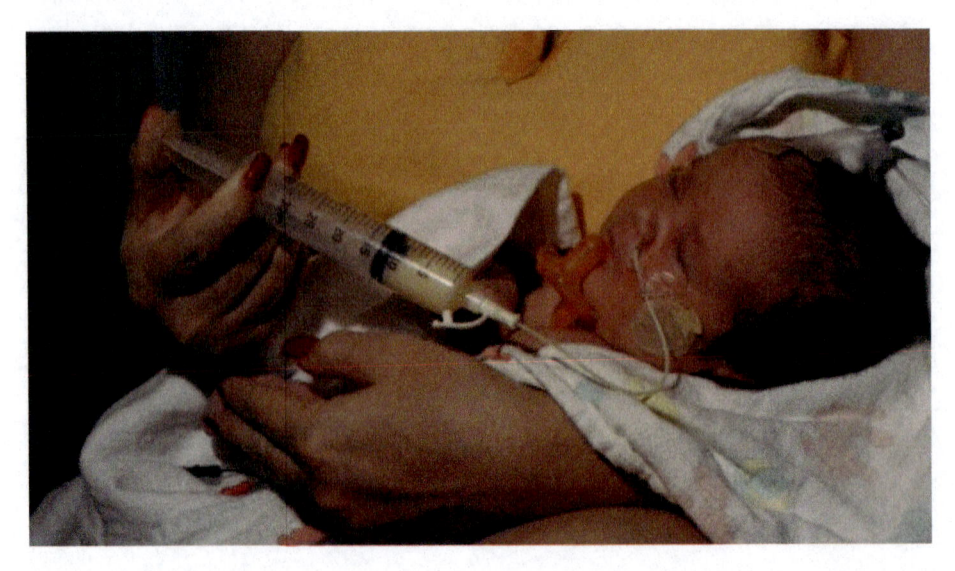

If a mother is not able to breastfeed her baby in NICU, it might be possible for her little one to receive carefully screened milk from a donor. This is the next best thing to mother's milk, and still helps to protect babies from complications including necrotising enterocolitis (inflamed intestines).

We know that frequent expressing, about every three hours, can help to promote breastmilk production. If a mother wishes to get to a situation where her baby receives exclusive breastmilk, she may have to work quite hard with the expressing until her baby is big and well enough to take the breastmilk directly at the breast. It might help to hire a hospital grade double electric breast pump when the mother is away from the hospital or has been discharged home.

I do not want my baby to have a bottle, what are my other options?

Babies can be offered milk through a feeding tube taped to a clean finger or the breast, a clean cup, spoon or syringe. Bottles are generally a sensible option for volumes over five millilitres and long term use.

Finger feeding is sometimes recommended for young babies who cannot breastfeed, or mothers who need to supplement their milk supply. Some hospitals can assist you by providing the tubes and showing you how to use them.

When finger feeding the adult puts a clean finger into the baby's mouth, nail side against the baby's tongue, and slides a thin flexible feeding tube (5 French Feeding or NG Tube) alongside their finger into the baby's mouth to deliver milk as the baby sucks. The tube leads to either a bottle or syringe containing milk. The flow of milk can be changed by raising or lowering the level of the bottle in relation to the baby. The higher you hold the bottle, the faster the milk will flow.

To be honest, finger feeding is not like breastfeeding, and a baby barely needs to open their mouth at all. The kit may be hard to keep clean and can easily become contaminated with bacteria. Finger feeding is rarely regarded as a feeding solution for more than a few days.

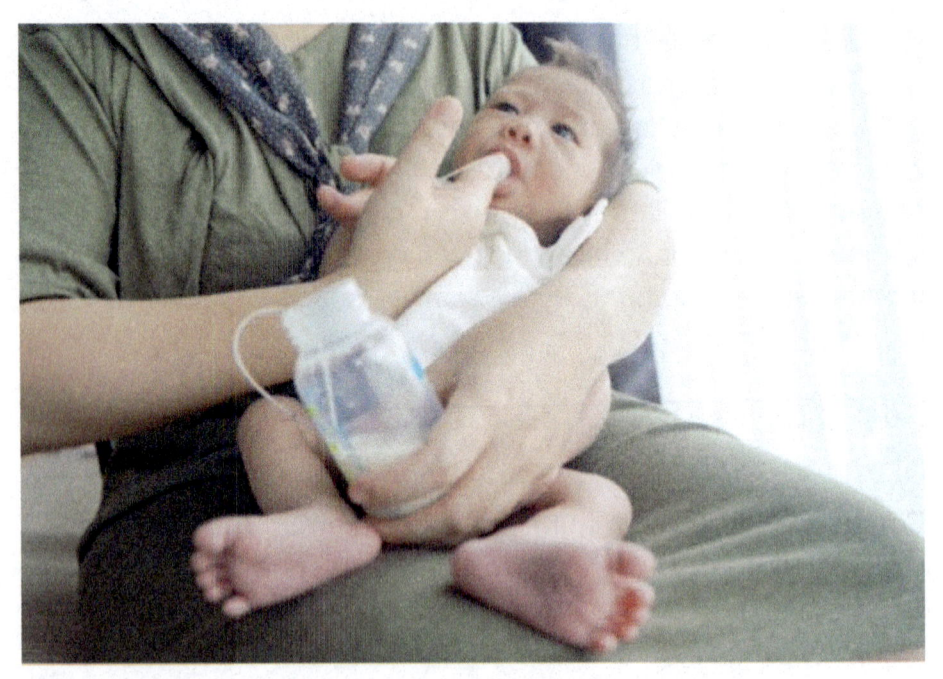

Supplemental nursing systems (SNS) or at breast supplementers (ABS) use a feeding tube taped to the breast near the nipple leading to a bottle of milk at the other end. Some feeding specialists might do similar by taping the tube to a nipple shield.

Hospital staff can often help you with using an SNS. You can also purchase a SNS on Amazon, there are several to choose from with very different price points.

These systems help people who are worried about their baby taking and then preferring a bottle. An SNS used at the breast can boost a woman's hormones and milk supply at the same time as providing additional milk. This can reduce time spent pumping.

Nursing systems are sometimes used by non-birth parents who are inducing lactation to stimulate milk supply, and/or for a bonding.

The tube is often taped on before the feed, but some people prefer to latch the baby on and then slide the tube into the side of baby's mouth.

Realistically, the tubes are sometimes hard to fit without someone else helping. They often slip or get pulled off. Cleaning them can be difficult. The manufacturers suggest flushing them through several times with hot soapy water and then with clean water. You push the water through with a with a 5 ml medicine syringe. Guidance varies on how long to keep each tube, anything from 24 hours to "a few days."

Cup feeding is another possibility, where you hold an open container to your baby's mouth and let them lap up the milk. This method is used around the world as cups are easy to get hold of and keep clean. You could use a special cup for the purpose from the maternity unit, or try an egg cup or shot glass. However, it is not the easiest thing to do without practice, please ask a midwife for help. It is important to do it properly and not pour milk into their mouth which could cause your baby to choke. Cup feeding is not like breastfeeding, and your baby is not learning sucking skills. Many parents find it messy and are not sure how much milk their little one has consumed rather than spilt. Most parents dealing with ongoing feeding problems ultimately find bottles more appropriate for longer term use.

Note that babies over four months old can use free flow cups from a supermarket or any suitable open cup. Given them lots of praise and cuddles as they learn to take milk for themselves. Much of the milk might be spilt at first. Somewhere closer to a year old, babies can try cups with an integrated weighted straw.

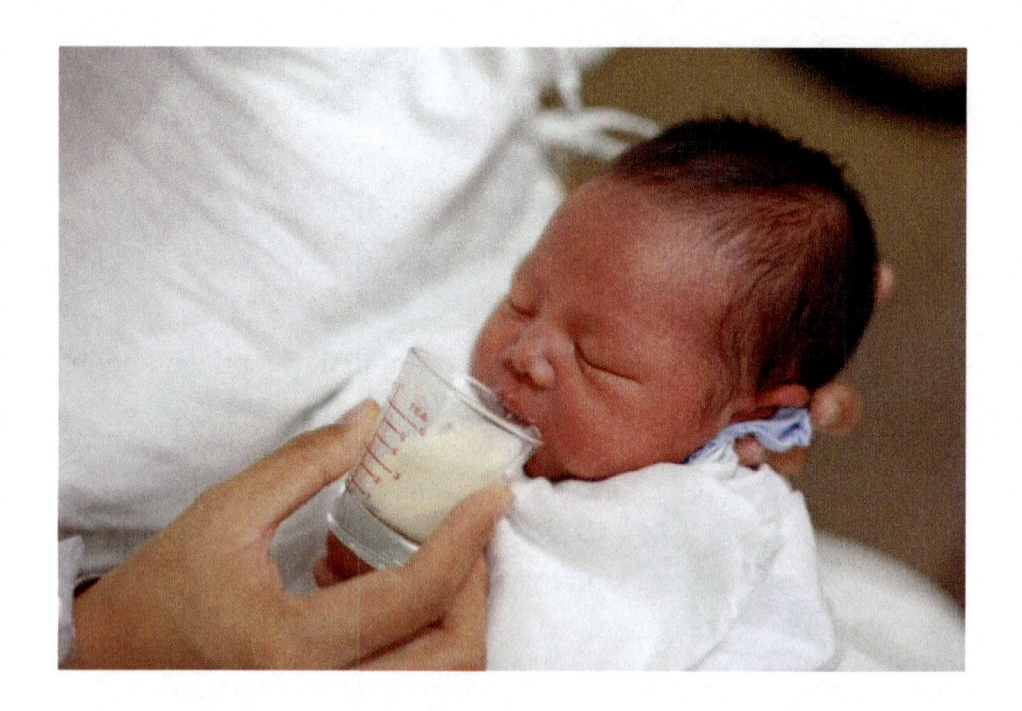

How to cup feed a young baby.

- Start with an appropriate small cup from the hospital, or something similar like a shot glass. Clean it thoroughly with hot soapy water and ideally sterilise it.
- Sit your baby upright on your lap.
- It often helps to swaddle your baby so their arms don't knock the cup. Wrapping them in a muslin would do, and it would catch any spilled milk.
- Hold the cup against your baby's bottom lip and tilt it so the milk is at the edge of the cup.
- The idea is for your baby to lap the milk, like a kitten, not for you to pour it into their mouth.
- Watch your baby carefully to see whether they want any more.

What is it like to spend time in hospital?
Most hospital staff want you to have a positive experience, but you may need to advocate for yourself or your partner.

I have spent a lot of my time on the postnatal wards at two busy London NHS teaching hospitals. While the midwives and staff are generally amazing and doing their best, there is no doubt that postnatal wards are often short-staffed, stretched, and not necessarily as efficient as you would like. You might find that people promise to bring you something, or come back with a test result or more information, and then you wait for hours. Shared wards can be very noisy with other families and their visitors, lights go on and off all night, staff come and make observations on you and your baby, and your exhaustion levels creep up.

Please be polite but assertive. If you are quiet with the curtains drawn around your bed, you are unlikely to receive much attention. It might be necessary to specifically ask for what you need. This is particularly the case if you are not mobile and need help picking up your little one. Each hospital will probably have qualified people who can help you with feeding in addition to the midwives. You might have to ask (several times) to see them. If you are in pain, or worried about something, please do not hesitate to ring the bell on your bed and ask to see your midwife.

If you receive particularly good or poor care, please note down the name of the staff members involved and feedback appropriately. NHS Hospitals have Patient Advice and Liaison Services (PALS), and you could also try writing to the Head of Midwifery at the unit.

Many women tell me that during this time they felt alone, detached, shattered, overwhelmed, frightened, not in love, tearful, smelly, violated, angry, let down, numb, grieving for the experience they wished they had. If this is resonating with you, I am so sorry.

Please reach out for support, and talk about it with friends and family. You might consider a hospital debrief service, or an organisation such as Traumatic Birth Recovery. If you are yet to give birth, do think ahead now about how you can look after yourself, and be looked after, in the postnatal period.

I missed the first hours of cuddling my baby. Have I ruined everything?

No! Please be kind to yourself, and cuddle your baby whenever you can.

It is lovely to have the "golden hour" of cuddle time with your new baby. We think that early breastfeeding or expressing in the first hour can help to boost long term milk supply. However, some women simply do not get the chance and it is nobody's fault. The good news is that bodies and babies are clever and will nearly always catch up given time, plenty of skin to skin contact, and frequent feeding or expressing.

It is normal to feel a sense of loss after a birth experience that was not your ideal. Take time to acknowledge this, talk about it, grieve for it. Then, think about what would feel lovely now and in the future. Possibly lots of cuddles in bed with your newborn, skin to skin. Maybe a soothing bath with your baby, with candles and some beautiful music, recreating a calm and wonderful space for you to connect.

Your partner might feel the same, and you could work on a gorgeous baby welcoming ceremony at home, or just find lots of times for cuddles in the days and weeks ahead.

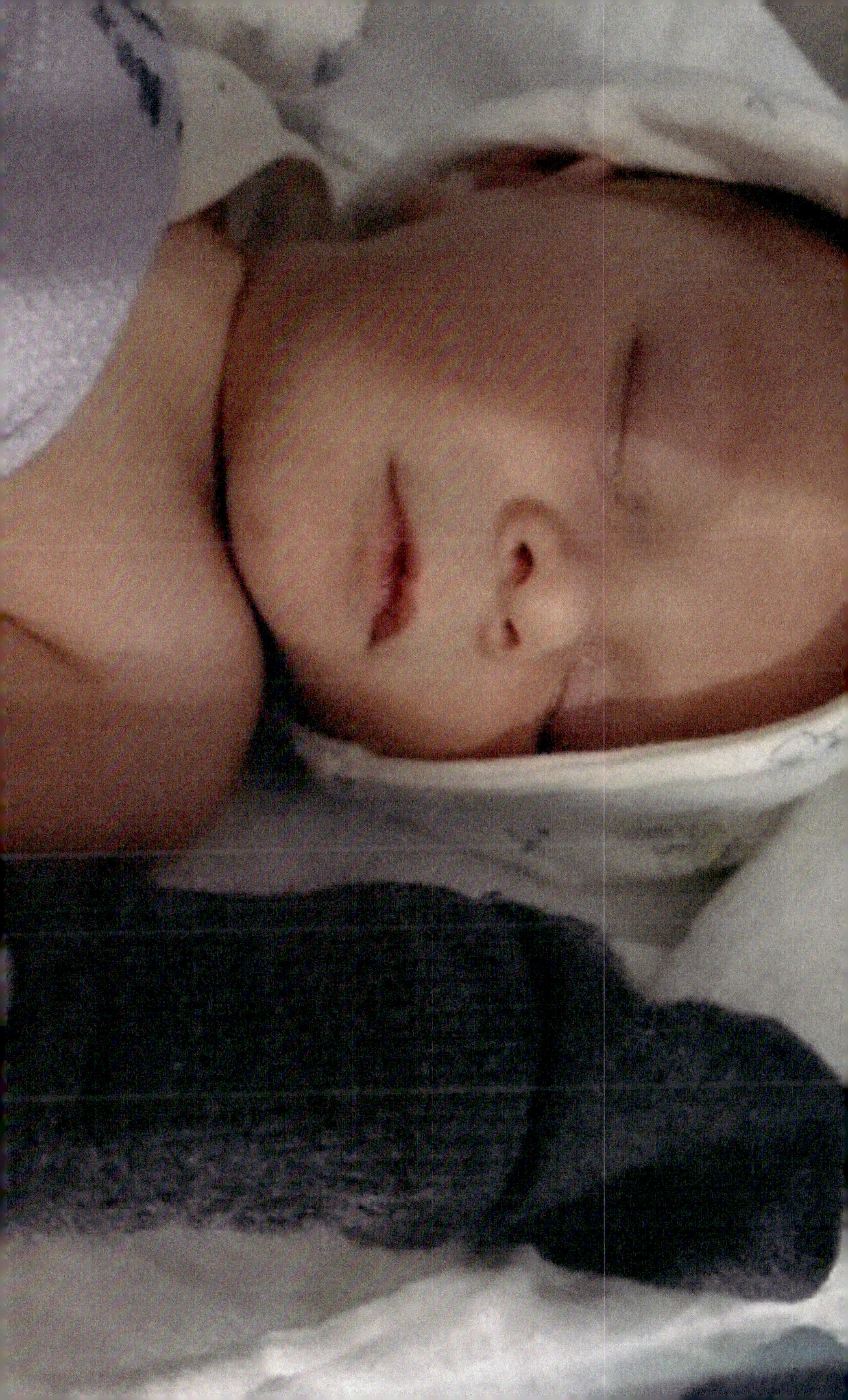

Someone gave my baby a bottle of formula. How will that impact breastfeeding?
It will probably be fine.

This is an issue that divides new mothers. I am equally likely to hear *"This amazing midwife came and took my baby away for a bottle of formula so that I could sleep"* and *"I can't believe it! The midwife gave my baby formula without asking me! She has ruined my chances of breastfeeding because now my baby will be confused, and I wanted to exclusively breastfeed!"*

There are a few things to unpick here. First, nobody should give formula without your consent. This is unethical and needs to be followed up with your healthcare provider.

Most midwives know that a new baby generally does not need formula unless medically indicated. This could be, for example, when the baby has low blood sugar (perhaps when mum has diabetes), the baby is jaundiced, the baby is clearly hungry and the mother is too unwell to feed, when the mother does not have sufficient colostrum right now because has lost more than average blood at birth or has retained placental fragments. In these situations, it will be essential for the baby to have food. If no mother or donor milk is available, this must be infant formula. In most cases, it is absolutely the right thing to give baby formula from a bottle and then get back to part or exclusive breastfeeding when everything is alright again.

In the meantime, if the mother is well enough, unrestricted skin to skin and breastfeeding or expressing will help to build the mother's milk supply as much as possible. If the mother is too unwell to do this, then her recovery has priority while her baby is fed with formula.

Some mothers are worried about their baby becoming confused by an artificial teat rather than a nipple. I have looked at the latest research on this, and it is inconclusive. A new baby's ability and willingness to suck on either a teat or a nipple is likely to be influenced by a huge range of factors.

Babies can have trouble sucking because of their:
- Lack of maturity.
- Feelings and pain levels after the birth.
- Sleepiness, which can be impacted by blood sugar, medications and levels of bilirubin.
- Mouth size and shape.
- Ability to move their tongue effectively.
- Sore or tight muscles from labour or birth.
- Birth injury e.g. forceps bruises.
- Ability to cope with the flow of the milk.
- Ability to cope with the shape and size of the nipple or teat.

Most of the time early feeding problems can be rectified with specialist help, perseverance, and patience. If the mum can protect her supply by expressing milk, I have seen babies get back to exclusive breastfeeding in a few hours, days or occasionally weeks.

Is my postnatal bleeding normal? Can it affect my breastfeeding?

Sudden profuse blood loss is a medical emergency needing immediate and urgent care in hospital and can affect your breastfeeding.

Your midwife will be very interested in the contents of your maternity towels! In the first 24 hours, you might need to change your maternity sanitary towels every one to two hours. After that the bleeding should calm down, and the blood gradually goes from red to brown and sometimes lighter coloured in the days and weeks after birth. It is normal to continue bleeding a bit for up to six weeks.

It is normal to pass some small clots, and to feel like you have a bit of flow when you stand up after sitting or lying down for a long time. If you do find a clot bigger than a 50p you can keep that maternity pad and show the midwife, or take a photo of it on your phone. If you pass a sizeable clot (like a golf ball or bigger), especially if it looks like it has tissue in it, please get someone to check you and it immediately in case it is left over placenta or a sign of infection.

You will need some big "granny style" black pants you don't care about and a lot of maternity sanitary towels, or buy adult nappy type pants. You are discouraged from using tampons or having sex while you are still bleeding postnatally. You might like to get some incontinence pads or disposable changing mats to sit on at home and to protect your mattress for a bit.

When is bleeding an emergency?

It is not normal to have sudden and profuse blood loss (like the feeling of waters breaking or peeing yourself.) If this happens in hospital, yell and push the emergency button. If it happens out of hospital, call an ambulance. This is an emergency.

It is not normal to have persistent increased blood loss. This is also an emergency.

It is not normal to have vaginal blood or discharge that smells offensive, or large clots especially with visible tissue in them. This could be an infection or retained products (bits of placenta) and needs to be urgently reviewed.

Significant blood loss, retained products or infection can all have a significant and sometimes long lasting impact on your milk supply, so please get some skilled support as soon as possible.

Can I breastfeed after severe blood loss?

If a woman has experienced blood loss of over a litre, especially just after the birth, it can have a profound effect on her milk supply.

Her body will have the colostrum that was already there, but the blood loss can interrupt the hormone production and transport that is necessary to stimulate further breastmilk production. It is quite literally an assault on the pituitary gland and the rest of her body. The blood loss and accompanying medications can leave a mother feeling weak, breathless and "out of it" for a few days.

Nobody in hospital wants to undermine a woman's confidence in her ability to feed her baby. However, this means that many women are not told about the likely impact of the blood loss. Sometimes their babies become hungry, dehydrated or lose a lot of weight in the first few days after birth. I feel it is much better to be honest. If the woman is well enough, she should aim for lots of skin to skin contact and breastfeeding or expressing. This should happen every three hours or more frequently if possible. If her baby is still hungry it is imperative that the baby is fed, and this can be with donor milk or formula milk.

The longer term impact of major blood loss varies enormously woman to woman. Some seem to get away with it, their milk comes in after a few days and they continue to breastfeed as normal. However, I have seen plenty of cases where that critical period means not enough breastmilk for this baby no matter how hard the woman works at feeding and expressing. If this applies to you, please see a breastfeeding professional through the NHS, a feeding organisation or privately, and work with them to figure out next steps. This might include topping up the baby with appropriate amounts of formula. There can undoubtedly be a period of grieving for the loss of any desired breastfeeding experience.

What is the breastfeeding impact of instrumental birth, or other birth injury?

It can be hard to move around and care for your baby, and your baby might also have a sore head or neck. Skilled supporters can help you to find comfortable positions to feed.

Many women will be very sore after a vaginal birth, especially those involving forceps or ventouse. They might have episiotomies or tears to their perineum or vagina, or into their back passage. These can be extremely tender for a couple of weeks. Haemorrhoids are also very painful for some women.

Women with damage to their vagina, labia, pelvic floor, perineum or bottom, will have a hard time sitting for long periods and also experience pain when moving around. There are special cushions which can help, or in desperation women often ask a partner or midwife to roll two equally sized towels into sausage shapes, put them down parallel to their thighs before they sit down, so that they do not sit directly on the sore areas. Feeding lying on the side might be hugely beneficial until the mother is healed.

It is appropriate to take some pain relief that is suitable for breastfeeding. Hospitals often prescribe dihydrocodeine, which is effective but can cause constipation so is not recommended beyond a few days. Paracetamol and ibuprofen are helpful. These medications are considered suitable for breastfeeding. Haemorrhoids can be treated with over the counter preparations like Anusol.

Babies can be sore after a difficult labour and birth and especially an instrumental delivery. Forceps can cause bruises or even cuts to a baby's face and scalp. Ventouse cups often leave rounded, raised bruises.

Babies with a sore head or neck often do not like being handled, especially forced onto the breast. They do best in laid-back or side lying feeding positions where there is less pressure on them.

Birth injury can put your little one at more risk of jaundice, which again makes feeding more challenging as they might want to sleep a lot. These babies often need to be woken to feed every three hours, from the start of one to the start of the next, until they are feeling better and parents are confident to feed them more responsively.

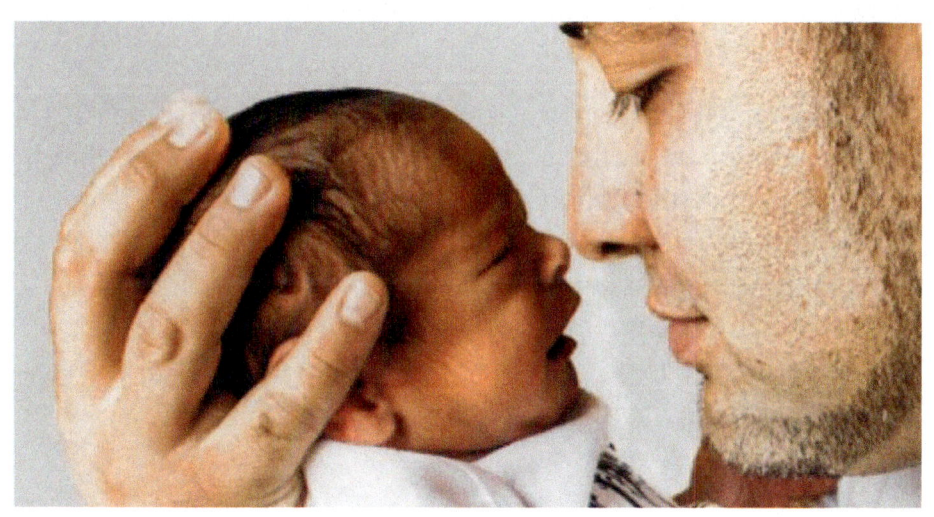

What are the challenges of breastfeeding after a caesarean?

The real difficulty after a caesarean is moving around and reaching for your baby. Twisting sideways to get your baby out of a cot next to your bed is next to impossible in the early days, and you will rely heavily on your partner or the midwives to help you.

It is ideal to begin breastfeeding in the operating theatre if you feel well enough, or in the recovery ward afterwards. If this is not possible, begin hand expressing colostrum to promote milk production within an hour of the birth.

Many women find the top of their bump feels very tender, and that they don't want the weight of the baby on it. They might prefer laid-back feeding positions in which mum is semi reclined and her baby rests tummy to tummy across her body. Alternatively, the rugby/football hold is useful.

Some mothers prefer to sit up with a pile of cushions on their lap, so that baby is supported without putting pressure on the bump or kicking the scar. Pillows also help mothers with cannulas in their hands, so that they don't have to hold the baby's weight along with everything else.

Caesarean births are associated with a slight delay in the milk coming in for some mothers. It is unclear whether this is related to stress, pain, higher blood loss, inability to move around, the mother or baby being unwell in the first place, or a combination of these factors. Most babies will be fine with their mother's colostrum in the meantime, but if the baby is particularly unsettled and hungry they can be offered some formula or donor milk as top ups while continuing to breastfeed.

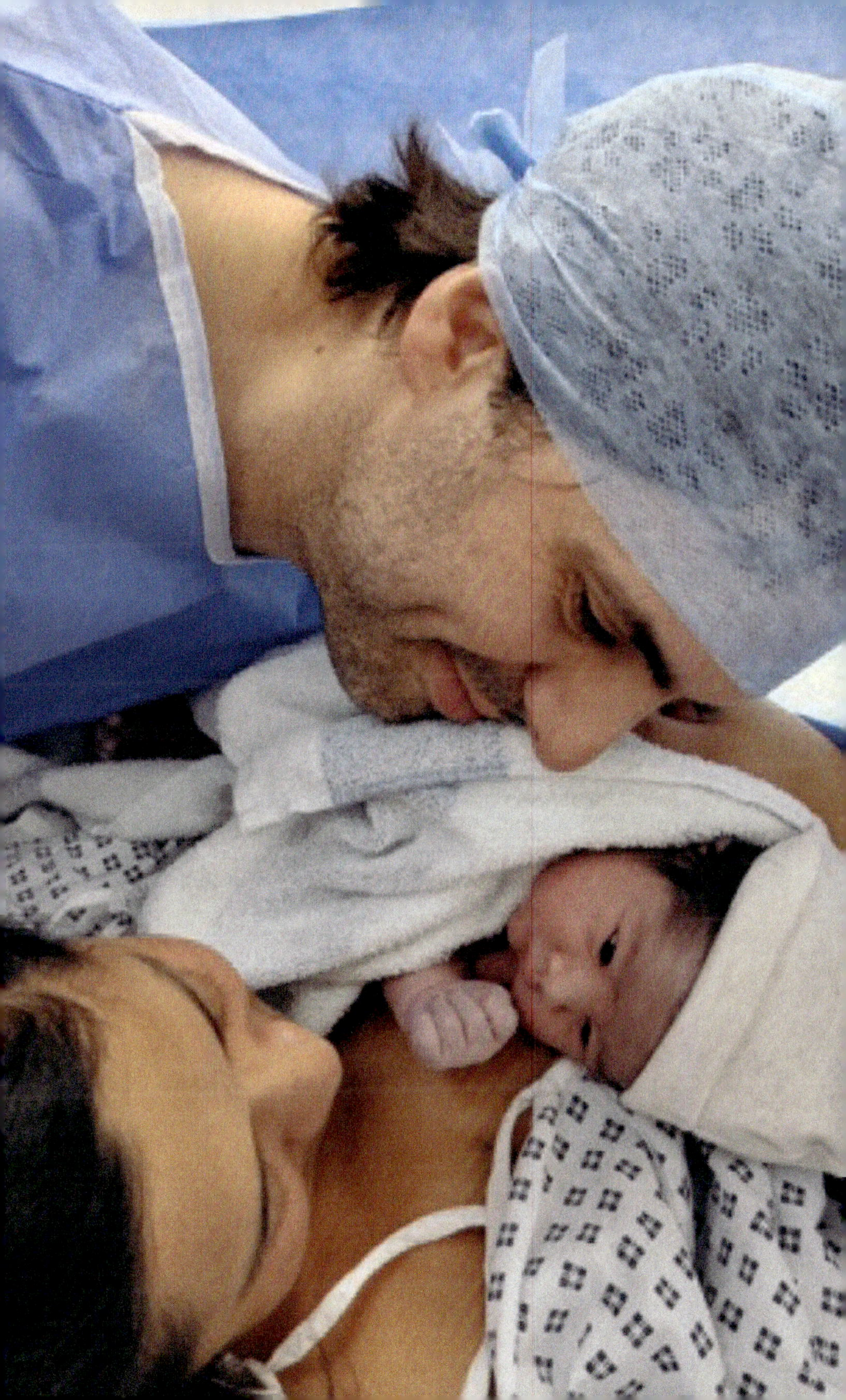

When shouldn't I breastfeed?

Some medications pass into breastmilk and can be dangerous for a baby.

This section can only provide a guideline. Please speak to your healthcare provider about all medications you are taking, or planning to take, including prescription drugs, over the counter medicines, vitamins and herbal therapies.

The Breastfeeding Network has evidence-based factsheets online and via their Facebook page.

You can search medications online at www.e-lactation.com.

Please consult with your healthcare team urgently before taking:

- Antiretroviral medications (for HIV/AIDS treatment).
- Cancer chemotherapy agents .
- Illegal drugs.
- Certain medications prescribed to treat migraines, such as ergot alkaloids.
- Mood stabilizers, such as lithium and lamotrigine.
- Sleep-aid medicines.

In addition, women who are undergoing radiation therapy should not breastfeed, although some therapies may require only a brief interruption of breastfeeding.

The advice around breastfeeding after general anaesthetic and certain diagnostic procedures is a little mixed and confusing. Some medical professionals suggest that mothers pump and discard their breastmilk for a certain time after a procedure, others suggest that the mother can breastfeed when she feels well enough. Please talk to your medical team about what they suggest in your case, it might depend on the medicines used.

This section can only provide a rough guideline. Please speak to your healthcare provider about all medications you are taking, or planning to take, including prescription drugs, over the counter medicines, vitamins and herbal therapies.

If you are worried that your baby is having a reaction to something in your milk, or is generally unwell, please seek medical help immediately. This could include excessive crying, excessive sleepiness, skin reactions, allergic reactions and diarrhoea.

Women with certain illnesses and infections may be advised not to breastfeed because of the danger of passing the illness or infection to the breastfed infant.

Breastfeeding your infant is not advised if you have any of the following conditions:
- Infection with HIV.
- Infection with human T-cell lymphotropic virus type I or type II.
- Untreated, active tuberculosis.

If you have access to safe formula milk you are generally advised not to breastfeed when HIV positive, as your baby has an increased chance of acquiring HIV through breastmilk. Antiretrovirals, when taken consistently, can reduce the viral load to undetectable levels. You may wish to consult with your doctors about your options for breastfeeding and weaning.

If you are sick with the flu, including the H1N1 flu (swine flu), you should continue to feed your baby. If possible, ask a well adult to look after your baby and feed your expressed milk from a bottle.

If you have Zika virus or live in areas where Zika virus is found, breastfeeding is still recommended. Although Zika virus has been detected in breastmilk, there are no reports of transmission of the virus through breastfeeding, and the benefits of breastfeeding are thought to outweigh the risks.

What is breastfeeding like for an autistic mother?

Autistic parents often face extra challenges in feeding their babies. They often report that medical staff are not well trained or understanding of their needs. It is hard to access skilled support. Advice can be confusing or contradictory. The unpredictability of new babies is very challenging when autistic parents need routines. Often the parents feel overwhelmed by all the physical contact.

It is estimated that 1-2% of the population are autistic. Adult women are less likely to be diagnosed than men, often because they present differently and are considered better at masking their autistic traits.

Autistic parents report challenges including:

- Lack of clear information and tailored face to face support for breastfeeding.

- Inconsistent and contradictory advice.

- Being touched without explicit consent, for example when being assisted to latch their baby on.

- Huge disruption to normal routines and ability to access coping mechanisms.

- Overwhelming feelings of being "touched out" from feeding and holding the baby.

- Unbearable sensations of breast or nipple pain alongside birth recovery.

- Feeling dirty, difficulties in maintaining hygiene in unfamiliar hospital settings and with a demanding baby at home.

- Sensory overload. Struggling with the noises, smells, food, others in close proximity and the unpredictability of a postnatal ward.

- Being told to respond to their baby's needs or cues, which can be hard to interpret and predict.

- Difficulties in accessing community support in a group setting, finding it hard to fit in socially and make new friends.

- Not being given definite times for appointments with midwives and health visitors at home, causing anxiety and disruption.

Many autistic women thoroughly research and prepare for breastfeeding. Some find the routines associated with making up and feeding bottles slightly more predictable and soothing.

In many cases, mixed feeding can allow the woman some time to rest and recuperate so she can breastfeed at other times.

Autistic partners face challenges too. Sometimes they find dealing with a baby very difficult, and the mothers are left with limited emotional and practical support. Often the woman will have to interpret what the baby needs for her partner. Autistic partners might do well with clearly defined roles and tasks.

This is an area that demands more research and attention. Autistic adults might seek peer support from trusted friends and family, and organisations like the National Autistic Society.

What is breastfeeding like after sexual abuse?

It is possible that you have been subject to past or ongoing abuse. If you are currently suffering abuse of any description, please seek help as a matter of urgency. There are many organisations that can help you.

Sexual abuse can have long term consequences that affect adults in many ways, and may be triggered by experiences of labour, birth and breastfeeding. In some cases, abused women cannot even contemplate the idea of breastfeeding, and that is to be respected. In others, breastfeeding a child can be a wonderfully healing experience. Some women choose to express their milk rather than breastfeed directly. If possible, let your care providers know what is going on for you so that they can be sensitive to your needs.

While this is by no means an exhaustive list, some women find it difficult to be observed or touched by health professionals, hold their baby skin to skin, feel baby's movements at the breast (especially in older nursing infants), cope with the sensation and look of the milk ejection reflex particularly if milk squirts out, and the physical sensation of milk on the hands or breasts. Some mothers may find it too much to feed their baby in bed especially at night, but could cope with daytime feeds. Many abuse survivors have learned to disassociate, and might find themselves feeling out of their body looking down at themselves with their baby.

Emotionally, abused women often suffer from low self-esteem. They may find it hard to ask for help, to accept help, and to build a support network of other adults around them. They are at increased risk of postnatal depression and anxiety. They are more likely to be in pain from other ailments than most women, including headaches and irritable bowel syndrome. These issues can result in parenting difficulties.

I very much hope that none of this applies to you, but if it does there are more resources in Sources of support.

Breastfeeding away from home

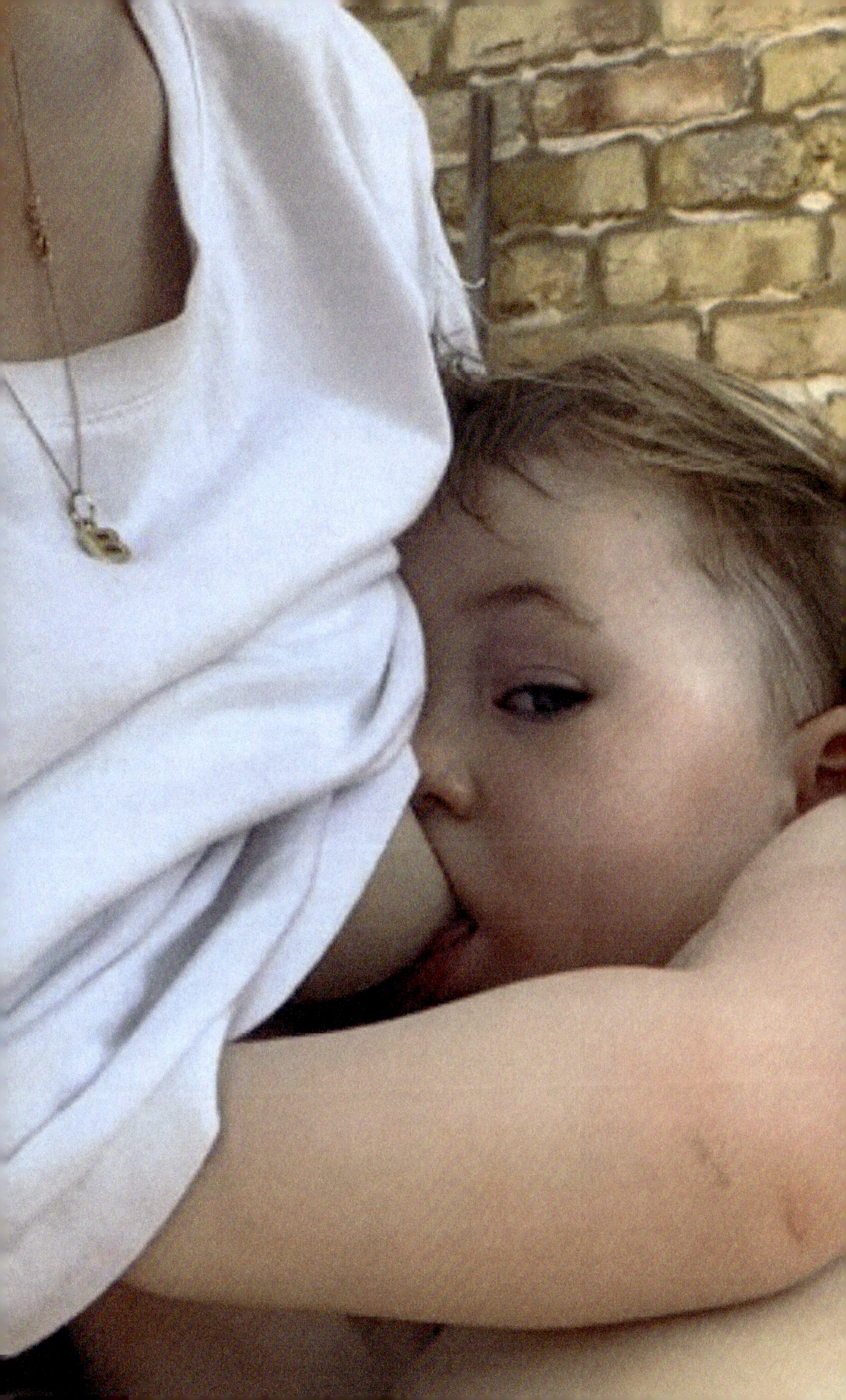

How can I breastfeed in public?

Breastfeeding in public is protected by law. You might like to be with other people who support you, especially as you practice the first few times.

Many women are apprehensive about breastfeeding in front of other people, including family and friends. This is perfectly understandable. Mothers might feel uncomfortable about others seeing an intimate part of their body, especially if their little one is hard to latch, or bobs on and off a lot, or when there is a lot of milk that might drip or spray out. Older babies tend to get distracted and look around and squirm when feeding. Some women use special items of clothing that are discrete and warm, others try breastfeeding aprons that hide what is going on underneath.

Lots of women are nervous about other people's reactions, especially when out and about, and may be concerned about where to find a breastfeeding friendly space where they can feed. It can be helpful to have a friend or family member with you when you first need to feed in public. You can look for breastfeeding friendly places, some of them display a sticker to show that you are especially welcome.

It is important to stress that **you are legally allowed to breastfeed your baby in public**. This might be a shop, café, park, museum, library, council office or on public transport. The Equalities Act makes it sexual discrimination to treat a woman unfavourably if she is breastfeeding. Try to remember that you are doing the most natural thing in the world, and the best thing for you and your baby. As society gets used to seeing breastfeeding mothers, the easier it will be for everyone.

Shall I cover my baby to protect my modesty while I am feeding?

If you want to.

If you feel more comfortable feeding with a shawl, blanket or purpose made breastfeeding cover, go ahead. You will still need to be able to see your baby to make sure they are OK. It is also important that your baby does not get too hot.

Can I feed my baby while they are in a sling?

Sometimes, if it suits your breasts and baby and you follow safety advice.

It may be possible to loosen off a sling and allow your baby to breastfeed. It depends on both the sling and where your nipples naturally hang. Your baby needs to be able to reach the nipple comfortably without you needing to move your breasts much. If you are using a sling, it is vital that you can always see your baby. Breastfeeding (or anything) in a voluminous sling where you can't see your baby is not safe. Check out the guidance on safe sling use.

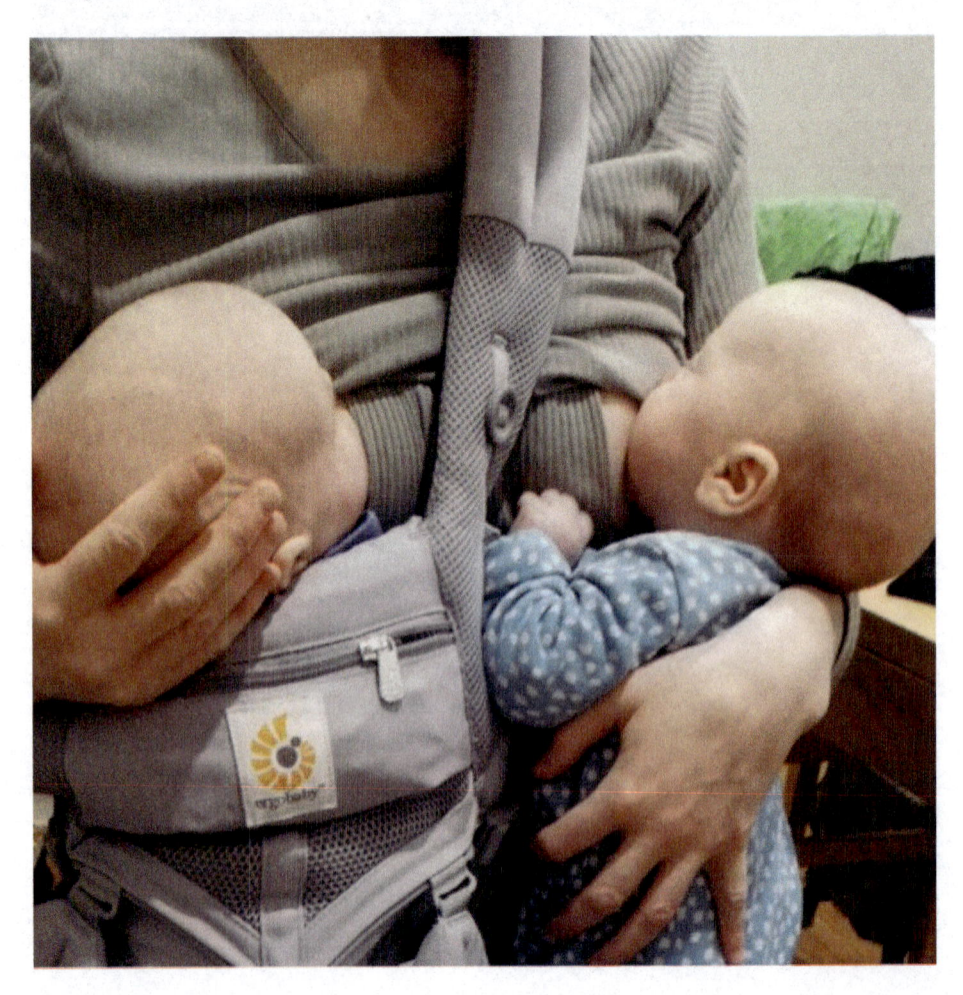

How can I breastfeed and return to work?

It is often possible to continue breastfeeding your baby after a return to work away from home. If you are employed, talk to your employer about breastfeeding before returning to work. Your next steps depend on the age of your baby and your breastfeeding goals.

Here are the questions that I get asked about returning to work.

Is my employer legally obliged to protect my breastfeeding or expressing?

Unfortunately, there is no specific law protecting breastfeeding or expressing at work in the UK. While it would be great for you both if your employer followed best practice and supported your feeding goals, they might not be willing or able to do so.

Employers are **not** legally required to provide somewhere for you to breastfeed a baby brought to you, or express and store your milk during the day, or to provide breastfeeding breaks. However, they **are required not to discriminate against breastfeeding mothers, and to meet health and safety requirements**. Guidelines suggest that women breastfeed or express in a comfortable, clean place with a lock, and this should not be the lavatory.

Will I have to express at work? How often?

Most women try to express milk at work at roughly the same time that their baby would feed, as this maintains their milk supply. It typically means expressing about every three hours. You might not have to express at work if your baby is older than nine-twelve months. You could feed just before leaving for work, and as soon as you are reunited, and not have to express in between. Every woman is different.

Some women arrange childcare near work so they can breastfeed during breaks.

Mothers are advised to keep a close eye on their supply, breast comfort, and their baby's nappy output. The pumping and feeding schedule might need to be adjusted according to how their body reacts. Luckily, breasts are quite resilient and can both raise and lower supply within a few days depending on demand.

How could I store and transport milk I have expressed at work?

There is a lot more on expressing and storing milk in the next chapter.

You could express your milk into a recently sterilised bottle, and store it ready for your return home. Ideally, you would have a cool bag with some ice packs or equivalent in it. Milk stored this way will keep for up to 24 hours.

Even better, put your stored breastmilk in the work fridge if available. Many women chose to put the whole cool bag into the fridge, so that their colleagues do not see or comment on obvious bottles of human milk. If the work fridge is under 4°C, milk will keep for about five days, but you must allow for the fact it will warm up slightly on the return journey so use common sense precautions. Keep it as cool as possible and use it as soon as you can.

Will my baby refuse to feed much in the day and then spend all night feeding instead?

This is a common concern. Women wonder how they will function at work if their baby feeds more frequently at night. It is likely that your baby will seek milk and connection with you when you are back from work, but not every baby does this.

It is important to remember that nearly all babies wake at night, and that this goes in peaks and troughs regardless of what is happening during the day. Many women return to work when their baby is nine months old and experiencing perfectly normal separation anxiety. They would be waking at night for comfort anyway.

You might like to think of strategies that get you all back to sleep as quickly as possible after disturbance. Breastfeeding is usually the fastest way to settle everyone. Co-sleeping with your little one can be a game changer.

Do I have to get my baby ready for my return to work and wind down breastfeeding in advance?

It seems a shame to stress both yourself and your baby by changing things in your last few weeks of maternity leave. It is often possible to maintain a close, ongoing breastfeeding relationship and not need to wind down breastfeeding as you return to work.

If you want to slow down or stop breastfeeding, most sources suggest dropping a feed every few days to allow your body and mind to get used to it.

Many women feel reassured if they have tried to get their baby used to taking a bottle or cup in advance of their winding down breastfeeding or having time apart.

What should I discuss with the people looking after my baby while I am at work?

Talk to the people who will be looking after your baby about what you would like to happen in various scenarios.

Try not to worry too much about the transition. Babies do seem to adapt to changes in routine and care. Many women are surprised about what their baby will do and eat in a childcare setting, that they would never do at home!

Discuss these with your childcare providers:

- Feeding on demand rather than schedule.
- Using paced/responsive bottle feeding.
- Not offering a full bottle within a certain time of you arriving home or at childcare to pick your baby up.
- Whether you would prefer them to prioritise solid food or breastmilk at mealtimes.
- Whether you are OK with your baby being offered formula if the expressed milk runs out.

In a childcare setting, establish:

- Their policy about receiving, labelling and storing your expressed breastmilk.
- How they would make sure that milk from different mothers was not mixed up in the fridge or given to the wrong baby (and what they would do if that happened).
- Where you can breastfeed on the premises if you want to when you are reunited with your baby.

Expressing and storing breastmilk

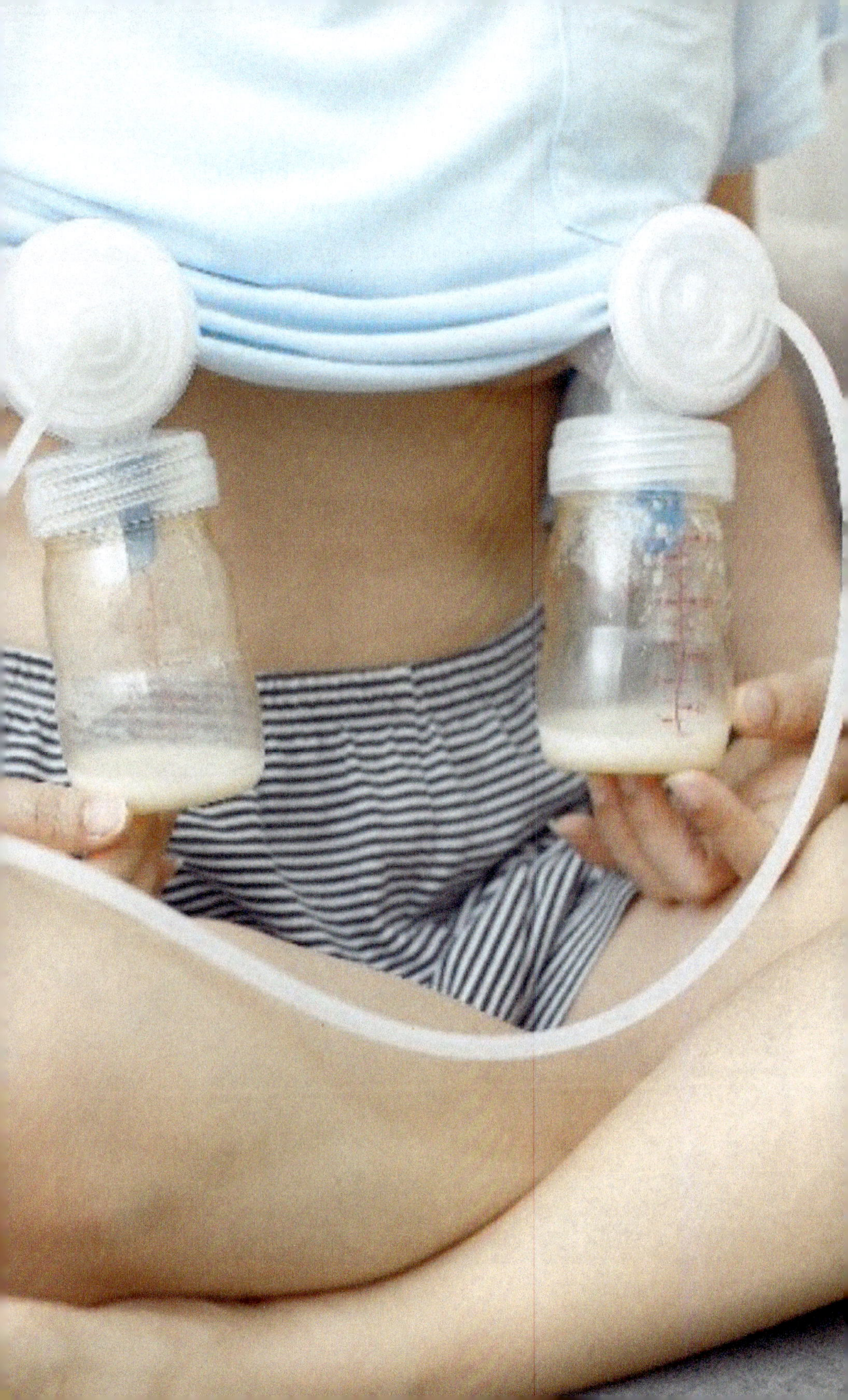

What is expressing?

Expressing means getting milk out of the breast by hand or machine to feed the baby or store until later.

This section covers the practicalities of expression and cuts through common myths and misconceptions.

Do I have to express?

No. But it can be helpful to boost supply, for babies who can't breastfeed directly, and to allow you time away from your baby.

While your social media feed might make you think that everyone is buying a breast pump, some women never want, or need, to express a drop.

However, many women choose to express milk for a wide range of reasons.

Many parents choose to feed their baby a bottle of expressed milk daily to give the mother a break and remind the baby how to take a bottle. Most of us have heard of babies that refuse to take bottles and wish to avoid this scenario, although there is precious little evidence for when to start or how frequently to give a bottle to avoid the problem.

Common reasons to express:

- Your baby cannot feed directly at the breast due to prematurity or illness or nipple difficulties.
- To relieve engorgement/full breasts when the milk comes in or at other times.
- To allow sore nipples time to heal.
- Needing or wanting to be away from your baby (occasionally or when returning to work).
- Boosting milk supply, especially if your baby does not breastfeed very efficiently.
- Getting your baby used to a bottle in the hope that your baby will happily switch between breast and bottle in future.
- Giving someone else the ability to feed your baby (hopefully while you rest).
- Offering family and friends the chance to feed your baby.
- When fully breastfeeding feels like too much, and combining it with bottle feeding is more sustainable.
- In instances of body dysphoria or trauma, when direct breastfeeding might feel bad.
- Building a milk supply if inducing lactation or relactating.
- Collecting milk to be stored for later, or for donation.

I hope this list has shown that there are many and varied reasons why someone might want or need to express milk. Expressing is by no means an easy option, or a cop out, or a failure. **It is a time consuming and loving act between the lactating person and the baby**. Expressing milk either short or long term can be a fabulous and often necessary part of a feeding solution. It often results in the baby receiving breastmilk for longer than they might have otherwise, with all the associated health benefits.

Do I need to "harvest" colostrum before my baby is born?

Many midwives suggest that women learn how to hand express and collect their colostrum from about 37 weeks pregnant. It is considered a useful and empowering skill to learn. There are pros and cons. Some women feel encouraged by seeing milk flow and others find it incredibly difficult.

Colostrum harvesting could be particularly useful for mothers with diabetes, breast or hormonal issues, or medical conditions that might affect the babies around the time of birth. A baby could be offered expressed colostrum to raise their blood sugar or supplement feeds while their mother works on maximising her milk supply.

The process is the same as for hand expressing, you can read about it in a few pages time.

If you manage to collect some precious drops into a feeding syringe, keep it in the fridge and try again a little later. At the end of the day, or if the syringe is full, put it into a clean plastic food container and pop it into the freezer. It will keep for six months. It can be thawed in the fridge for a few hours or at room temperature, or just hold the syringe in your hand for a bit.

Please do not get disheartened if you do not see any colostrum. The placenta is busy telling your body that you are still pregnant. What you gather before birth is not a good indication of the amount of milk you will produce afterwards.

Shall I express by hand, manual or electric pump?

Hand expressing is the only sensible way to get colostrum out. After that, it is up to you.

Colostrum is thicker than more mature breastmilk and is made in small volumes. Breast pumps are not good at extracting it and it tends to get lost around the valves in the pump. Most mothers hand express instead, which involves squeezing the area just back from the nipple and collecting the colostrum in a milk collection syringe, sterile teaspoon or other sterile utensil.

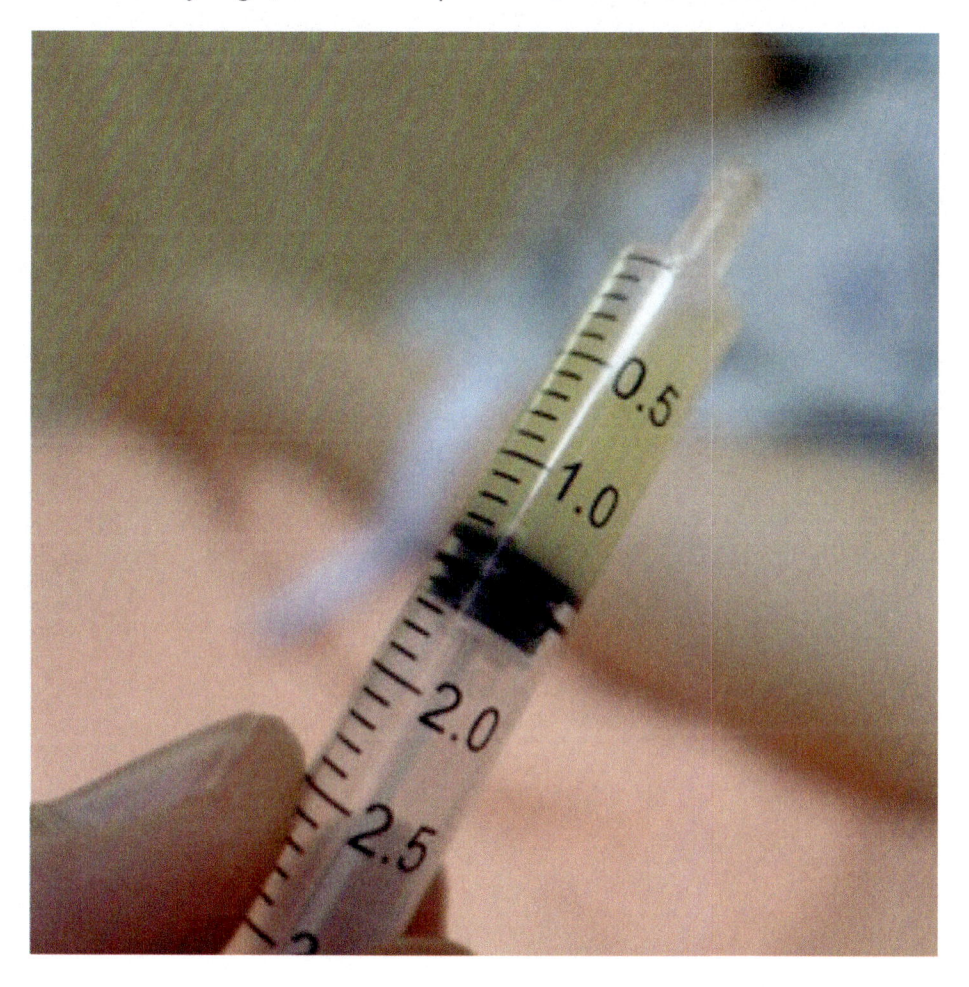

After the milk has come in, mothers can use **silicone collection systems**. These are not pumps, they just create suction on the nipple. The idea is to feed or pump on one side and simultaneously use the device on the other to collect milk with minimal effort.

On the plus side, these are cheap and effective. However, you need to be also feeding or pumping one side while using them to collect milk from the other. You must have sufficient supply to both feed and collect extra during one feed cycle. For this reason, many mothers use them in the mornings or other times when their milk supply tends to be higher.

Hand pumps have valves and a sucking action much like electric pumps. They are discreet and relatively inexpensive, and have less of a "dairy" feel about them! However, daily and repeated use of hand pumps can be hard work on the wrists.

Domestic electric pumps are available in single and double versions, depending on what you plan to use them for. They are not all created equal, so do check online reviews for recommendations. Electric pumps tend to be quite efficient provided they are working and fitted correctly.

Wearable pumps are newer additions to the market. They often charge up like a phone and tend to be controlled via an app. You wear them in your bra, although some still have motors that stick out and would show under your clothes. These pumps are generally quieter and more discrete than other electric pumps, and women appreciate the ability to move around. Some wearable pumps still have tubes connecting them to a machine, so double check what you are buying.

The shape of wearable pumps means that they are not always ideal for women whose nipples naturally point downwards or for those with larger breasts in general.

Feedback suggests that in some models the motors are not as strong as some of the standard electric pumps. Mothers who are exclusively or extensively pumping milk in the first month or so might benefit from a pump with a stronger motor.

Rental grade pumps are bulky and expensive to buy, but an excellent choice to hire if you find yourself needing to express for every feed. For mothers of multiples, babies in NICU or those exclusively expressing, hiring a highly effective pump would often make life much easier. They are available direct from the manufacturer or through some larger pharmacies.

When should I start expressing?

This depends on why you are expressing. If your baby has just been born and cannot latch on, you will need to hand express. After your milk comes in, you can use an electric pump.

If you are at home and worried about milk supply, you could express at any point to reassure yourself that your baby has had something, and to build up supply. Expressing might be necessary to relieve primary engorgement when the milk comes in.

It is important to remember that leaving the breasts full for long periods (more than four to five hours) can reduce your milk supply. On the other hand, expressing much more than your baby needs will boost your milk supply and can result in discomfort.

For mothers whose breastfeeding is going fine and there is no urgent need to express, it is common sense to wait until you and your baby are used to breastfeeding before adding expressing into the mix.

Somewhere between four and six weeks is often suggested as a good time to start, as most babies are amenable to taking bottles at this stage.

How do I express by hand?

Hand expressing is a skill that can take some practice. However, once you find a way that works for you it is cheap and effective.

How to express by hand:

- First, wash your hands with warm, soapy water, and have a sterile container or feeding syringe ready to collect the milk.

- You might like to have your baby close to you, or a picture of your little one, or something that smells like your baby such as a blanket.

- Some mothers find gently massaging their breasts before expressing helps their milk to let down.

- Cup your breast with one hand then, with your other hand, form a "C" shape with your forefinger and thumb. Squeeze gently. It shouldn't hurt.

- Where to squeeze varies from woman to woman. I suggest starting around the place where the coloured part of the areola meets the fleshy part, and work backwards and forwards from there until you find the sweet spot for you. Squeezing on the "sticking out" part of the nipple, if you have one, does not work.

- Release the pressure, then repeat, building up a rhythm. Try not to slide your fingers over the skin.

- Drops should start to appear, and then your milk usually starts to flow.

- If no drops appear, try moving your finger and thumb slightly. Once you have been expressing one area for a while the flow will slow down, then move your hands around and try another spot on the breast.

- When the flow from one breast has slowed, swap to the other breast. Keep changing breasts until your milk drips very slowly or stops altogether.

How do I feed my baby expressed milk in the first few days?

You could feed your baby using a feeding syringe angled into your baby's cheek, or use a cup or bottle.

Feed your baby when they are awake. Rouse them if necessary, by picking them up or putting them down, removing some of their clothes, opening their nappy, or putting a cool wash cloth on their body. These early feeds are often a two person job!

To use a feeding syringe, hold your little one securely and offer them your clean finger to suck with the nail side down. Slide the syringe down the side of your finger into your baby's cheek. Squeeze in a small amount, say 0.1 ml at a time, and wait for your baby to swallow before gently giving them some more.

Slightly bigger volumes can be given in a sterile cup or a bottle that is suitable for a newborn.

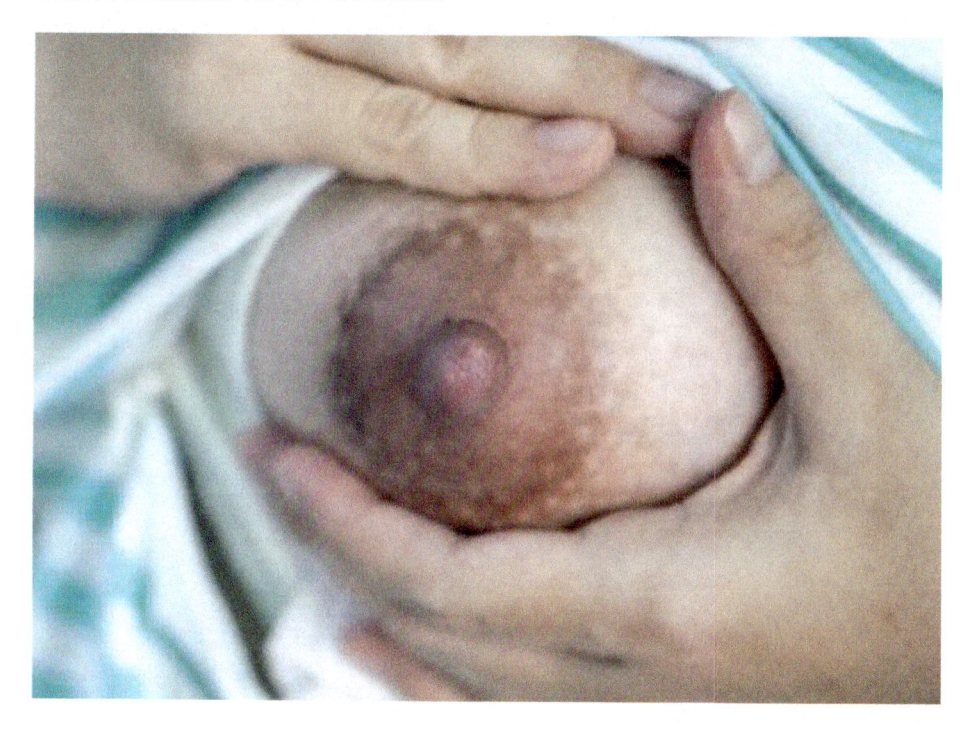

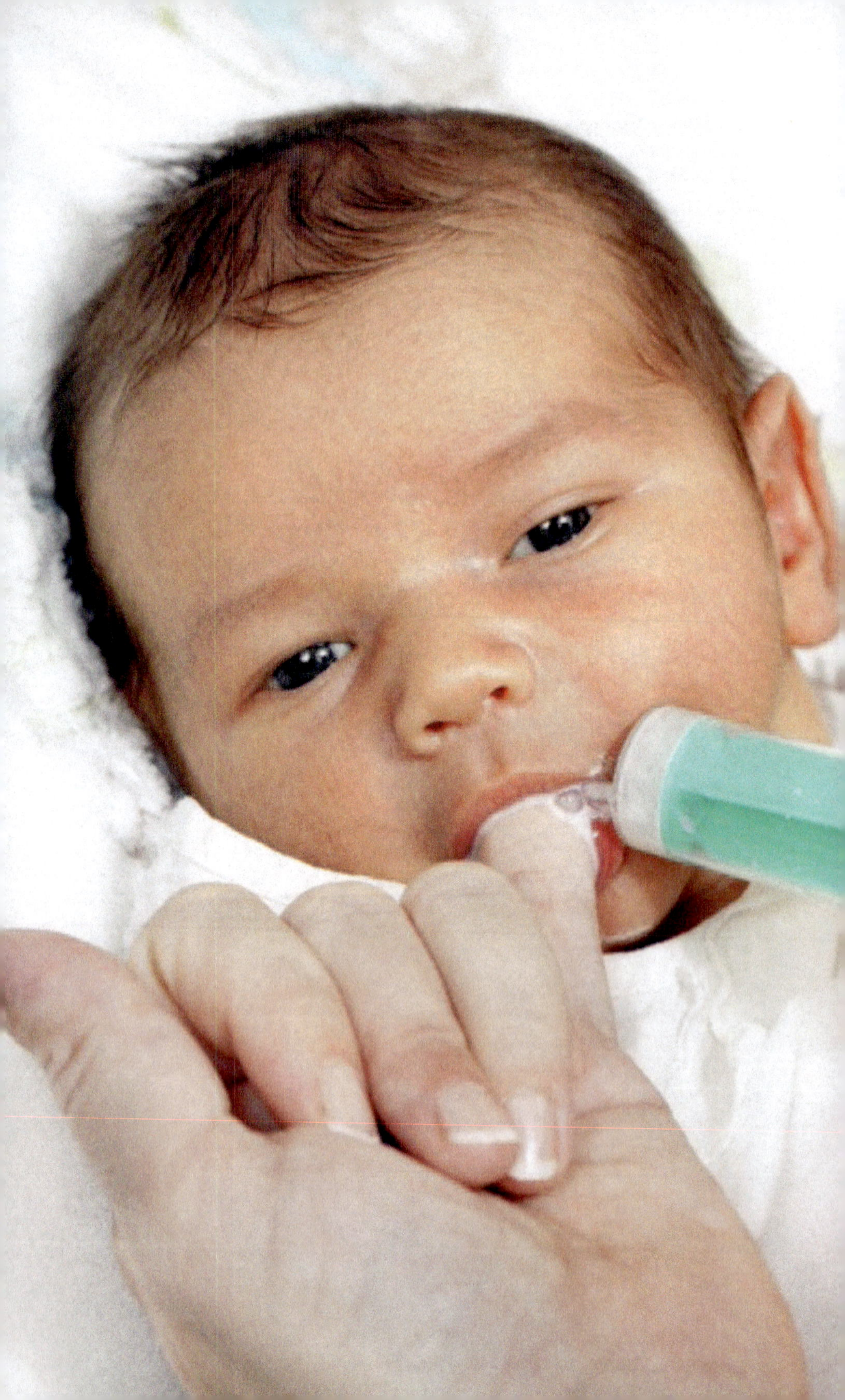

How long shall I express for each time?

Why are you expressing? If you want to ease engorgement, just pump until you are no longer painfully full. If aiming for a large volume of milk, pump until milk is no longer actively coming out and then maybe even a few minutes more.

Every woman's let down rate is different. Some can pump as much as they want or need in five minutes, others might take 10-20 mins. Excessively long sessions of 30 mins or more might cause some breast soreness. We know that pumping for shorter times more often is more effective in boosting milk supply than one or two marathon expressing sessions. Hand expressing before and after pumping maximises the output.

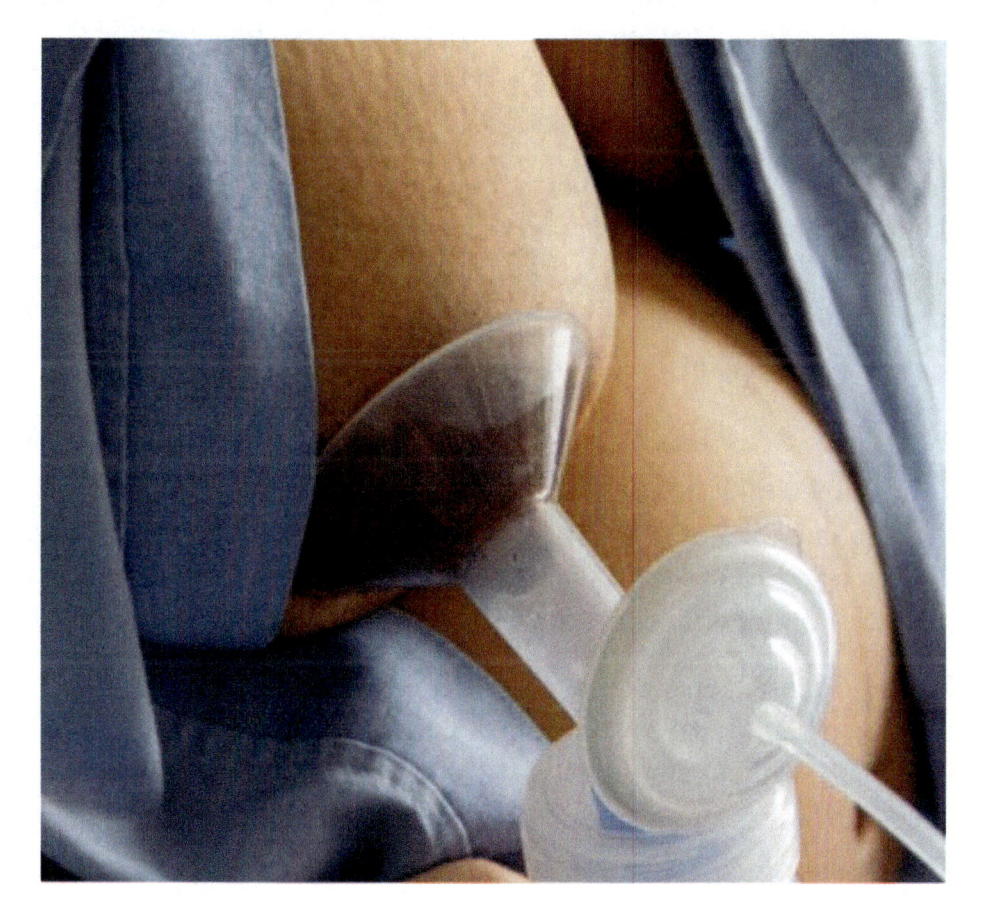

What time of day should I express?

When you pump will be determined by what you are expressing for. There is generally a bit more milk available after midnight and in the mornings.

If you are away from your baby at your normal feed time, you may wish to express for your comfort, to maintain your milk supply, and perhaps you can store the milk to give to your baby at another time.

If you are about to go out without your little one, or go to bed for a rest while someone else feeds, you might express just in advance in order to help your comfort and store milk for the baby.

Parents wanting an evening bottle will often try expressing after a morning feed to collect any spare milk, and then express both sides before they go to bed, with the hope that this together generates a full feed for the partner to give. Everyone is different.

You might have heard that your breastmilk varies between day and night and that milk expressed in the daytime might keep your baby awake if given at night. I have not found any evidence for this. Most newborns eat and sleep in the daytime just as well, or perhaps better, than they do at night!

How much milk should I express for a feed?

Your baby's milk intake will vary enormously depending on the age and size of your baby, number of feeds each day, how hungry your baby is at that moment, how hot it is, and how well they take breast and bottle.

Try to be led by your baby as much as possible. Aim to have enough in your bottle to know that your baby has had all they need. If they drain it and frantically look around for more, you will need to offer additional milk. If they drink well and leave a tiny bit, that sounds about right.

If you are keen to know volumes, try 90-120 ml for a baby over four weeks and see how you get on. Younger babies may not need that much, and newborns much less.

Please ask your midwife if you are unsure about feeding your newborn, and look out for other signs such as their behaviour and what is in their nappy.

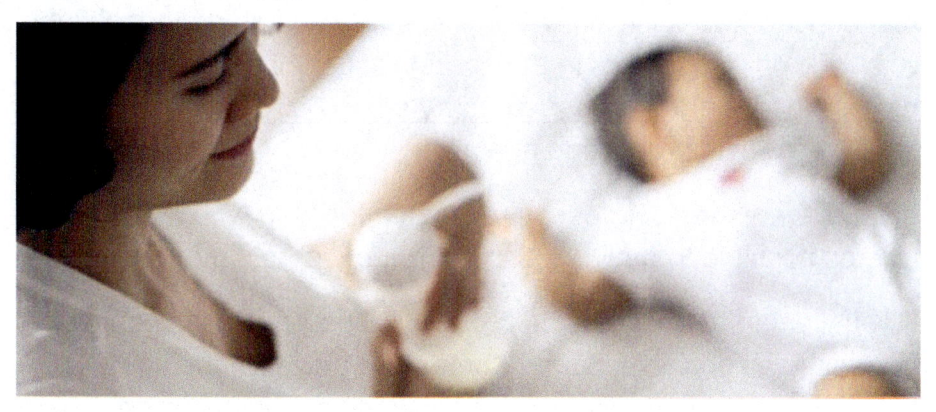

One side or both?

This depends on why you are expressing.

If you are apart from your baby you might like to try both sides to minimise time and maximise milk production. If you want to boost your milk supply, do both. If you are concerned about oversupply or trying to wind down breastfeeding, you might do the minimum expressing necessary to maintain comfort.

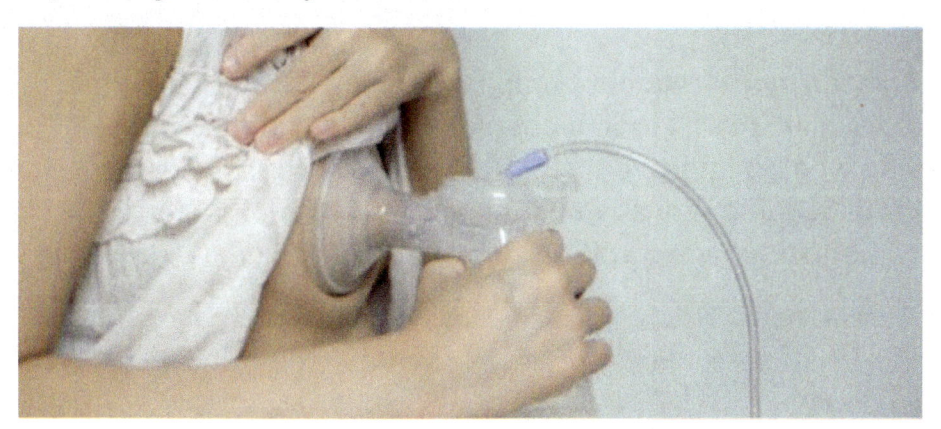

What impact will expressing have on my milk supply?

The emptier the breasts are, the more milk they produce. Frequent expressing encourages the body to make more milk.

Breastfeeding works by supply and demand. Each breast responds to how full it is. As the breast fills and starts to feel tight with milk, little receptors in the breast tissue tell it to stop making more. When the breast is drained by a baby or by expressing, it starts more milk production again.

If you were expressing because you felt full, you would take off as little as possible to avoid asking your body for lots more milk.

If you wanted to boost supply, you would be advised to pump as frequently as possible. Some women try "power pumping," that is multiple short expressing sessions each day for a period. This signals to the body that much more milk is needed.

Pumping milk and offering it later the same day gives the breasts the same amount of stimulation overall. It is unlikely to have a big impact on milk supply unless the mother leaves her breasts uncomfortably full for long periods.

How do I maximise milk production when expressing?

Shorter, more frequent pumping sessions tend to yield more milk overall than a couple of long sessions. Women who want to improve their supply, or are exclusively breastfeeding, are advised to express six to eight times in 24 hours including at least once at night.

If you're not with your baby, try looking at a photo or snuggling with your baby's blanket to get the oxytocin going and the milk to let down.

Start an expressing session with a gentle breast massage. Hand expressing before and after feeds maximises the output. You can boost the amount collected by using breast compressions. Put your fingers and thumb either side of the breast and gently squeeze together about once per second. You are not wringing the breasts, siding your fingers towards your nipple or causing any pain!

Pumping both sides at once, or feeding on one side and pumping the other, saves time and boosts production through the oxytocin produced.

There are certain foods, herbal supplements and medicines that may help, but none will be as effective as frequently taking the milk out.

Stress and anxiety can slow the let down of milk. Many women find that covering the collection bottle with a baby sock or similar, miraculously reduces their performance anxiety.

The excellent book "*Making More Milk*" 2nd Edition by Marasco and West, has chapter and verse on everything to do with milk production.

How many times a day do I need to express?

Mothers who are exclusively expressing will find that they have a magic number of times they need to express to maintain their supply.

It is based on lots of unique factors including breast storage capacity. Some women express everything their baby needs in a day in three sessions, others might need up to eight.

Why is nothing coming out?
There are a number of things to consider, most of them fixable.

First, please check that the pump is working and on the correct setting.

I know this is basic, but you'd be surprised what I see. The sneaky little valves get broken easily. Check the suction. Increase the suction to where your nipples become sensitive and then back off. We do not want expressing injuries. Some pumps have a variety of funnel sizes, and you might want to double check that yours fits well. They should be effective at getting milk out, and not cause sensitivity, swelling or pain.

Next, ask yourself whether anything is impacting on your supply, especially if you are in the first couple of days after birth. Some temporary issues, and some longer term medical ones, can impact your milk production. If you are concerned enlist some skilled support.

Some women can fully breastfeed their baby and never express a drop, it just happens. Reducing performance anxiety, for example by covering the bottle with a baby sock, helps some mothers.

Some mothers find that one day nothing comes out, even if they have been exclusively expressing for ages, but it is often better the next day. If the problem persists please call a lactation professional for advice.

Will giving my baby bottles of expressed milk confuse them and ruin breastfeeding?

We do not have enough scientific evidence to answer this question. If your baby needs expressed milk, please give it to them, and get some skilled support. Most young babies take to both breast and bottle.

If your baby cannot breastfeed and you want to give your breastmilk, expressing milk is your only option! It has the dual benefits of feeding your baby and maintaining your supply until a time when your baby can (hopefully) breastfeed directly. Some midwives recommend using a cup, or feeding tube taped onto your finger or nipple, rather than giving your baby a teat. However, this can be impractical in the longer term.

In my experience, most young babies take willingly to both breast and bottle. However, we know that some babies (and parents) develop a preference for a bottle, especially when breastfeeding has been difficult or stressful. Please get some specialist advice. I have seen women quickly lose confidence in breastfeeding when their baby seems unsettled at the breast and a bottle becomes an easier option. Lots of skin to skin contact and breastfeeding practice in a low stress environment (not when your baby is really hungry) can help. This might mean starting with a bottle and finishing the feed at the breast.

We can help your baby by giving bottle feeds in a slow and paced way, using a slow flow teat if possible, to make the experience more like a breastfeed.

There might be times when baby or mum's anatomy or some other factor comes into play and bottles are by far the most sensible way forward.

Pumping and dumping – is it necessary?

Rarely. There are some medical conditions, investigations and treatments that make breastfeeding unsafe for a length of time, so mothers express and discard milk.

We all want to keep our babies safe and protect them from harmful substances in our breastmilk. So, when is it necessary to express milk and throw it away?

Right up front, I don't like the phrase "pump and dump." Expressing milk is an act of love from a mother for her baby, and discarding breastmilk can feel soul destroying.

The good news is that pumping and throwing away breastmilk is only necessary on rare occasions.

What if I am away from my baby and expressing for comfort?

This is probably the most common scenario. You are out for the night or at work, and your breasts feel uncomfortably full. You need to pump to make yourself feel better and to retain milk supply. Of course, if you have access to a clean and sterilised pump and a fridge or freezer bag you could store this milk. If not, then discard the milk and carry on with whatever you were doing.

What about when I have been drinking alcohol?

Guidelines suggest that you can drink small amounts of alcohol occasionally without needing to pump and throw away the milk. A limited amount of alcohol makes its way into your breastmilk, but as your blood alcohol level falls the alcohol also leaves your breastmilk.

The key thing is that you can't look after your baby properly if you are drunk! A sober adult needs to do that until you are feeling better. If you are drunk, it is probably safer to pump and dump if your breasts feel full. It would be dangerous to feed your baby in this state, and you should not bedshare or risk falling asleep on the sofa with your little one.

What if I smoke cigarettes?

Breastfeeding is still your best option and there is no evidence to suggest for pumping and dumping after smoking cigarettes. You might find factsheets from the Breastfeeding Network helpful.

What if I take recreational drugs?

It is vital that you do not use recreational drugs or street drugs while breastfeeding. If you have taken them or plan to do so, please contact the National Breastfeeding Helpline or your healthcare provider about how to keep you and your baby safe. If you have taken cocaine, it is essential that you do not breastfeed for 48-72 hours, so please express and discard milk over this time. Methadone that has been prescribed appears to be safe.

What if I am unwell?

If you are ill with a cough, cold, stomach upset, fever, flu, mastitis or other common illness, it is rarely necessary to pump and dump. The illness will not be transmitted through the breastmilk, and chances are your baby was exposed before you even knew you were sick. Your milk will contain protective antibodies that actively protect your little one. The closeness and protection of breastfeeding will almost always be the best thing for the baby.

Women with certain illnesses and infections may be advised not to breastfeed because of the danger of passing the illness or infection to the breastfed infant.

Breastfeeding your infant is NOT advised if you have any of the following conditions:

- Infection with HIV.
- Infection with human T-cell lymphotropic virus type I or type II.
- Untreated, active tuberculosis.

What if I have been diagnosed with nipple thrush?

You can carry on breastfeeding while both you and your baby are treated for the thrush symptoms. You will be advised to discard previously expressed milk containing thrush from your fridge or freezer.

What if my nipples are bleeding?

Please get some help with this! The milk is fine for your baby, but can look alarming for the mother or if baby spits some up. The main thing here is to sort out the causes of your sore nipples!

What if I have herpes outbreak or cold sore?

Herpes can be very dangerous or even life threatening for a newborn baby. Herpes virus infections create blisters or sores on the skin, which might affect the breast or nipples or present as a cold sore. It is imperative to consult your doctor for a diagnosis and the latest advice.

If you have active herpes sores, be meticulous with hygiene, washing your hands frequently. Prevent your baby from touching the sores until they have dried. Research suggests that you can breastfeed if you cover the sores so that your baby does not touch them. If you cannot do this on one or either breast, you will need to pump and discard your milk while baby is fed through a bottle with uncontaminated milk or formula. Throw away your milk if it, or the pump parts, have touched the herpes sore.

If you have a cold sore on your lip, you can breastfeed but do not kiss your baby and please be super cautious with hygiene.

What if I need to take some medications?

Please speak to your healthcare provider about all medications you are taking, or planning to take, including prescription drugs, over the counter medicines, vitamins and herbal therapies. Pain relieving drugs like ibuprofen and paracetamol are compatible with breastfeeding. Most antibiotics are also fine. The Breastfeeding Network has evidence-based factsheets online. Most drugs have not been tested on lactating mothers and clinicians may be cautious. There is more detail on medications in the Special situations chapter of this guide.

How often should I sterilise expressing kit?

The NHS recommends that you thoroughly wash up and sterilise everything that touches your milk before expressing e.g. collection containers, pump parts, flanges.

The guidance on sterilising varies so much, it can be very confusing. It is important to wash your hands and the pump parts in hot soapy water before beginning an expressing session.

The good news is that breastmilk has natural antimicrobial properties. For this reason, several pump manufacturers and many hospitals just ask women to wash it all in hot soapy water in between expressing sessions and sterilise the kit once a day. Some women who are expressing frequently do not wash up between every single session, and instead put all the parts in a food safe bag or plastic container in the fridge. There is more about methods of sterilising later in this book.

How do I store expressed breastmilk?
Apply good food hygiene standards to stored breastmilk.

You can store breast milk in a sterilised container or in special breast milk storage bag:

- For about four hours at room temperature.
- In the fridge for up to five days at 4C or lower (you can buy cheap fridge thermometers online). Put it in the back of the fridge, not near or in the door.
- For two weeks in the ice compartment of a fridge.
- For up to six months in a freezer.

The storage guidelines vary. While expressed milk is naturally antibacterial, you still want to be careful with it. If you are worried that the milk smells "off" please do not give it to your baby.

Ideally do not add warm newly expressed milk to cold milk already in your fridge; if you want to combine them for later wait until the new batch is cold too.

Breastmilk that has been cooled in the fridge can be carried in a cool bag with ice packs for up to 24 hours.

Storing breastmilk in small quantities will help to avoid waste. If you're freezing it, make sure you label and date it first.

Expressed breastmilk may naturally split out with the cream on top. Just give it a swirl before giving it to your baby.

Sometimes pre-frozen breastmilk smells funny. If it consistently smells soapy you might find it helps to scald the milk before freezing it.

Is frozen milk the same quality as freshly expressed milk?

Freezing breastmilk causes some physical changes to its components, but the milk is still good for your baby and preferable to formula.

There are not many large studies detailing the impact of refrigerating, freezing and rewarming breastmilk. Results suggest that it preferable to refrigerate rather than freeze the milk for short term storage of 3-5 days. It is great to use any stored milk as soon as possible.

Milk frozen for more than two weeks will have reduced fat and protein content, and less vitamin C than when first expressed. The bio-active proteins in breastmilk are also affected by handling and storage to some extent. However, previously frozen breastmilk is still great and better for your baby's health than formula milk.

How do I defrost frozen breastmilk?

Breastmilk that has been frozen is still good for your baby and retains many of its amazing natural properties. It is best to defrost frozen milk slowly in the fridge before giving it to your baby.

If you need to use frozen milk straightaway, you can defrost it by putting it in a bowl or jug of warm water or holding it under running warm water. It is not a good idea to microwave it. Once it is defrosted use it as soon as possible and within 24 hours. Do not re-freeze milk that has been defrosted.

What about warming breastmilk?

You can feed expressed milk straight from the fridge if your baby is happy to drink it cold. Or you can warm the milk to body temperature by putting the bottle or bag in a jug of warm water or holding it under running warm water.

Do not use a microwave to heat up or defrost breastmilk. Overheating could damage some of the important proteins and other constituents in breastmilk. It could also cause hot spots, which can burn your baby's mouth.

Do not tip out the milk and warm it in a pan on the stove.

Once your baby has been drinking from bottle of breastmilk it should be used within about two hours and anything left over must be thrown away. We don't have much evidence around bacterial growth in breastmilk, and it depends on a lot of factors such as overall hygiene and storage conditions. The two hour suggestion is a precaution.

How can I express for my baby in Neonatal Intensive Care (NICU) or Special Care Baby Unit (SCBU)?

Hospitals have kit to help you express and safely store your milk, and can advise you on how to do this at home. Your baby will receive the milk by cup, bottle or NG tube into their mouth or tummy until they are well enough to breastfeed.

It is a wonderful gift to express milk for your baby while they are growing and recovering. Expressing boosts and maintains your milk supply. Breastmilk is the best possible thing for your sick or early baby. Ask the medical staff in NICU/SCBU for lots of opportunity for skin to skin contact with your little one if it is possible, it will help you both to feel better and will promote the conditions for breastfeeding.

Babies are typically able to suck and swallow milk from about 34 weeks gestation, although obviously this is assessed on a case by case basis. Premature babies look perfect but miniature. However, they often do not have the energy, cheek strength or even mouth size to exclusively breastfeed.

In the early days, your baby might be able to lick milk off the nipple if you squeeze it out. Hours, days or weeks later they might be able to latch on. NICUs sometimes suggest nipple shields to help with this. As your baby grows you will find they can take more and more direct from the breast. You will be encouraged to work at breastfeeding and your baby might also receive breastmilk, donor milk or top up formula. This can be given through a feeding syringe, bottle, cup, or flexible NG tube taped to your breast or direct into their tummy.

Ideally, you would express milk at least six times a day. Expressing during shorter, frequent sessions is more effective than one or two marathons. You can boost the amounts considerably by hand expressing just before and after you pump.

Ask the hospital staff caring for your baby for advice on how to express and store milk. Units tend to have slightly different protocols and equipment. Hospitals have amazing breast pumps and sterile collection kits. Your milk will be properly labelled by the nurses or midwives and stored in the milk fridge until your baby needs it. Once you get home, you can hire hospital grade pumps from some pharmacies or direct from the manufacturer. Medela and Ardo are recognized brands. Visit the Bliss and UNICEF websites for advice on expressing milk for a premature or sick baby.

An expressing regime can be hugely challenging, both practically and emotionally. Some women find it gives them something meaningful and useful to do to help their baby. Others find it overwhelming, especially if they are also unwell. I hope that you have lots of loving support, because the physical and emotional toll of all this can be enormous.

It is easy to become disheartened if your milk supply cannot keep up with your baby's demand. Please remember that any colostrum or breastmilk is beneficial for your baby. It is thought that one millilitre of colostrum or five millilitres of milk contain 1-5 million white blood cells that help your baby's immunity. What a wonderful gift to your baby! Any expressed milk is to be celebrated.

Many hospitals may offer carefully screened donor milk to sick or premature babies with your permission. This is protective for their sensitive digestive tract, and could be a fabulous option if your milk is not available in sufficient quantities.

How can I exclusively pump milk for my baby?

Some incredible mothers do not feed their baby at the breast, but express milk to be fed by bottle. This is a huge undertaking and testament to how much a mother will do for her child.

Exclusive pumping can be necessary or desired for many reasons including:

- Separation of mother and baby.
- Health issues in mother or baby.
- Latch or milk transfer issues that are impossible to resolve.
- Preference of a parent who does not want direct sucking on the breasts.
- When a child is in daycare or being looked after by someone else.
- Inducing lactation or relactation.
- Pumping for another person's baby, or to donate milk.

What kit do I need?

Exclusive pumping takes time, and it is great to buy or rent the best possible double electric breast pump that best suits the needs of the parent. Wearable breast pumps allow women more chance to do other things alongside pumping.

How many times a day do I pump?

Breasts make milk in response to stimulation and milk removal. Pumping mothers often find that they need to pump multiple times a day. Gold standard would be 6-8 times, which would include at least one session overnight.

After a few weeks of pumping, a mother will be able to assess how often she needs to remove milk from her breasts to maintain the supply that she desires or her baby needs. This frequency varies from woman to woman. I have worked with exclusively pumping women who can get a full day's supply with three pumps, and others who would need six to eight pumps a day. The number of pumps partly depends on the milk storage capacity of the breasts.

Please work with a lactation professional if you are worried about how many pumping sessions are right for you. If you feel supply going down, you might need to adjust your pump times, kit or schedule until you find a combination that works.

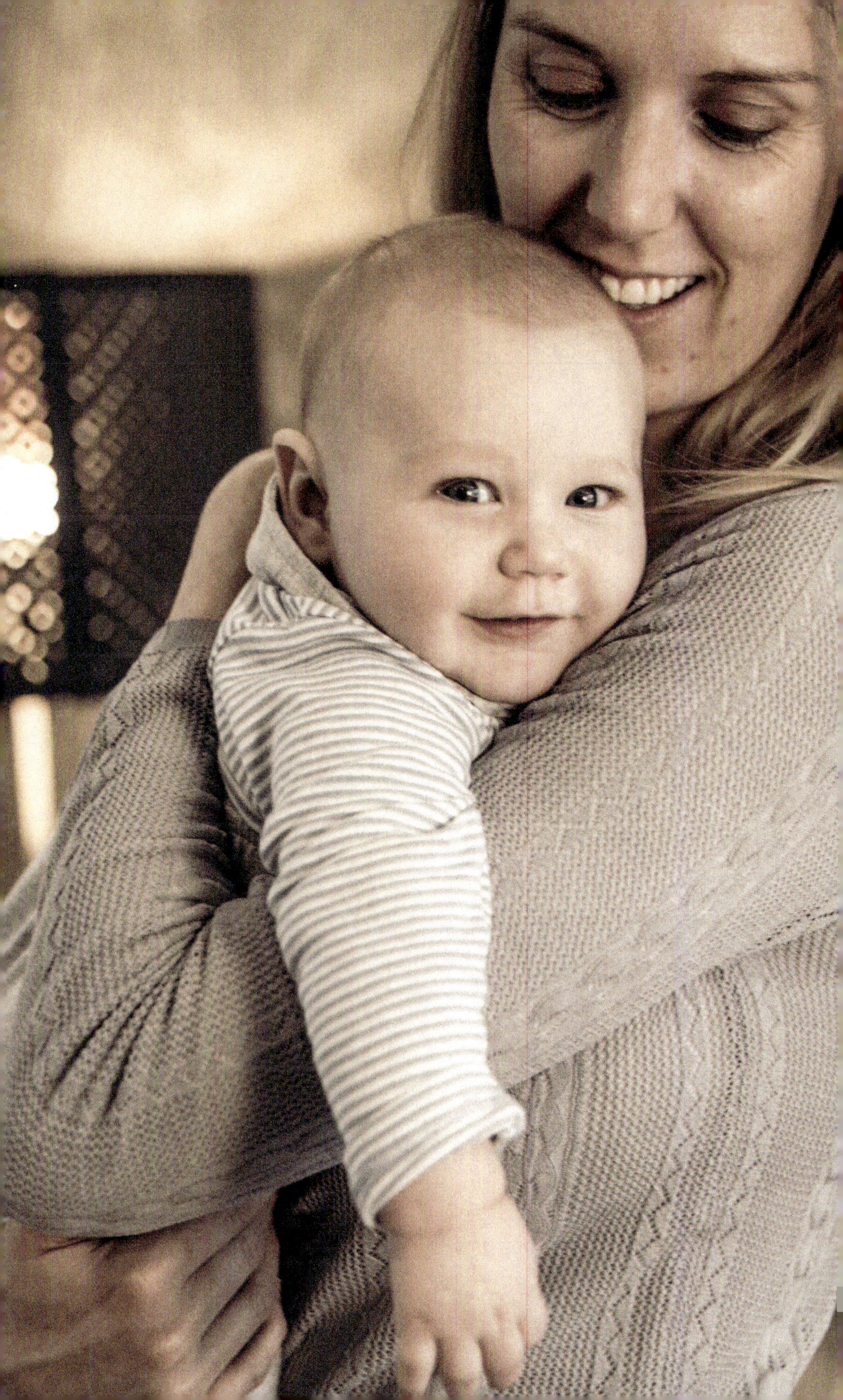

Bottle feeding basics

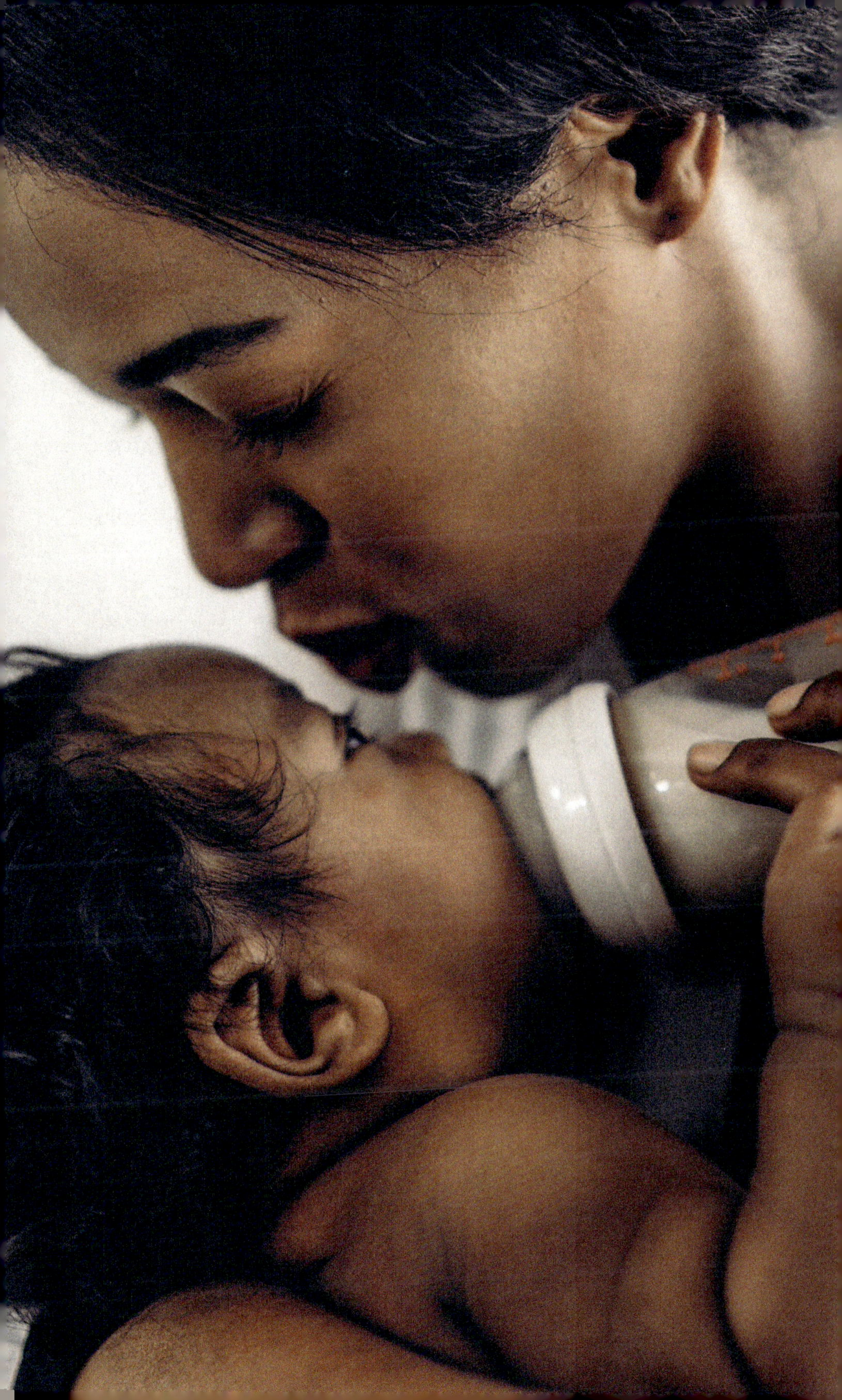

Which bottles and teats are best?

Pick a bottle and teat flow rate suitable for the age of your baby. You might need to experiment to find the optimal bottle for your little one. Conical teats tend to be well tolerated by many babies.

New parents often worry about what to buy for their baby, and there is an almost overwhelming amount of choice out there. Each baby bottle claims to be closer to breastfeeding, to prevent colic or wind, to keep your baby comfortable. Should you choose a latex or silicone teat? What about the flow rate? Are the expensive bottles worth it? You are unlikely to find any independent research into the marketing claims, but your family, friends and social media all have clear opinions.

The bottom line is that most babies cope with most bottles. Buy one or two bottles at first and see which one your baby prefers. However, you may have to experiment to find the one that your baby can use most effectively.

Tips for choosing bottles:

- Although many bottles come in small and large size, there is no benefit to buying small ones and swapping later.
- Wide necked bottles can be easier to fill, especially if using formula powder.
- Some people like to use the bottle that fits with other equipment such as their breast pump and steriliser.
- Anti-colic bottles have multiple parts and can be tricky to clean well. They are also more expensive. To date, there is no independent research that they help reduce colic! Common sense would suggest trying a more basic bottle first.
- Glass bottles are recyclable, chemically stable, and reduce plastic waste. However, they are also heavy and breakable.
- Some bottles have features that might appeal to you, such as the option for sterilising the bottle by itself in the microwave.

Tips for choosing teats:

- It is important to find a teat that your baby can take easily and not too fast, so that full bottle feeds generally last somewhere between 10 and 20 minutes. As a suggestion, it might take your newborn about five minutes to drink 30ml. **Many brands offer a slow flow teat for newborns and then medium and faster flows for older babies**. There is a considerable difference between the flow rates in different brands.
- Some teats are variflow or only offer milk when your baby sucks well on them. This is great for some babies, but can be difficult or impossible for babies with a compromised or weak suck.
- Orthodontic teats work well for some babies, but others are less able to get milk from them. We don't have any firm evidence that babies need them.
- Latex teats are often softer than silicone and can be a good choice for babies with a weak suck such as premature babies. They are slightly less durable than silicone teats, and latex can occasionally cause an allergic reaction in some people.
- Your baby needs to be able to grasp the teat deeply with an extended tongue and be able to maintain a good seal with relaxed lips. If your baby is squirming, readjusting, sliding up and down or leaking milk from their mouth during feeds, they might need an alternative teat. A teat shape that slopes gradually from tip to base (such as the Lansinoh Natural Wave) can help many babies to keep a wide open mouth.
- Tiny or premature babies and those with suck problems might need a narrower base.
- Teats vary in firmness and texture. Babies with tight, tense mouths or those who collapse teats while feeding will need a firmer brand. Some women aim to match the shape and texture of their own nipples so that the teat feels more familiar. Remember that your nipple will be a very different shape while your baby is actively sucking on it!

What does a good latch look like?

The picture below shows a baby feeding well at the bottle, the baby has relaxed lips that are turned outwards on the bottle teat.

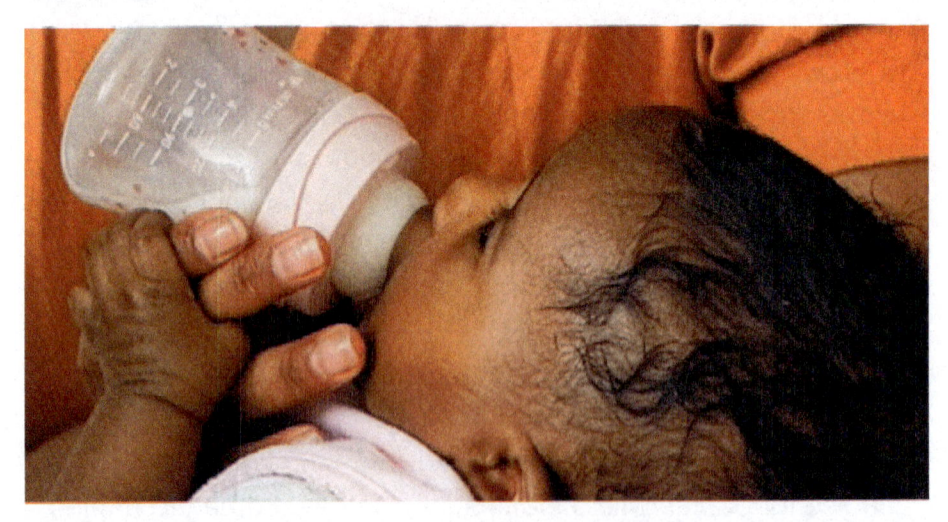

A baby with a good seal on the bottle will drink comfortably at a pace of about a suck a second, with some pauses between bursts of sucking. They will not make slurping sounds while feeding, leak milk from the sides of their milk, or repeatedly let go of the bottle.

In contrast, the following baby is not bottle feeding well. The baby has a very shallow latch on the bottle, and their lips are rolled in. This baby might benefit from a different selection of bottle teat.

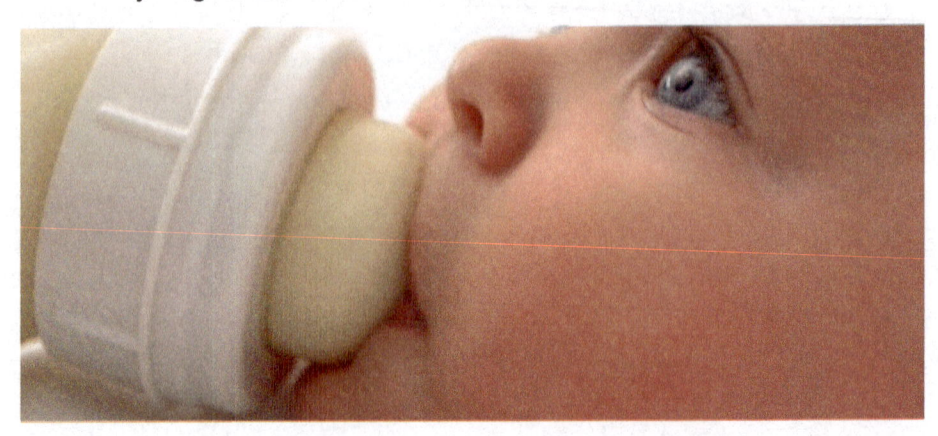

How to choose infant formula.

A standard first formula labelled 1 of any brand will suit most babies. Ready to feed formula is considered safest for your baby in the first two months.

Many parents find anxiety levels rising when confronted with the huge choice of formula milks available.

The key point is that newborn infants need either liquid or powdered formula that is **suitable from birth**, which will be **labelled 1**. This is appropriate all through your baby's first year, after which your baby can be offered full fat cow's milk instead.

Marketing claims on formula packets can be misleading. The basic contents are heavily regulated. Additional components in "premium" brands may not be better for your baby. The research indicates that the cheaper brands are just as good as the high price brands.

Pre-prepared formula is generally convenient and has the advantage of being sterile until is it opened. The World Health Organisation suggests ready to feed formula is the safest option for babies under two months. Formula feeding parents will be asked to bring clean and sterilised bottles and ready-made liquid formula into hospital.

Powdered formula is a more cost-effective option than ready mixed formula. **It is important to make the formula correctly using boiled fresh tap water that is still over 70°C, and then cool the bottle for the baby.** The issue is bacteria in the formula powder, not the water. Recent university-led research suggests that most parents do not make infant formula correctly, and that bottle preparation machines may not be heating water sufficiently to kill bacteria in the formula powder. This is of particular concern when babies are young or premature, or unwell.

Formulas labelled "follow on formula," "hungry baby," "goodnight formula," or "lactose free" are **not** recommended. In many cases they are an inferior product to standard number 1 formula, even though they are often more expensive. Parents are advised not to use specialist formula such as lactose free or soy. If a baby needs specialist formula in the UK it will be prescribed by a doctor.

The organisation First Steps Nutrition Trust has conducted independent research into infant formula. Their reports are free to download from their website. I have summarised their information in this guide.

Is organic formula better?

We have little evidence that organic formula is better for your baby's health.

Several brands such as Kendamil and Hipp offer formula made with organic milk. While there is little evidence that this is significantly better for your baby, it might sit well with your own ethics on farming and animal welfare standards.

Is Comfort formula good?

There is no convincing evidence to suggest that these milks are effective in preventing colic, wind, gastrointestinal discomfort, or constipation.

Several well known formula brands offer a "comfort" milk. Babies do not need the hydrolysed proteins they contain to make infant formula easier to digest.

Comfort milks play on parents' fears and experience of babies with "tummy troubles." It is normal for newborn babies to have wind and instances of gastric discomfort, and to have unsettled periods in the evenings. When babies fuss it is often called colic, but this is not a recognised medical diagnosis.

Several recent reviews have reported that there is not enough evidence to use any specialist formula for the relief of colic, and the NHS advice relies entirely on practical and soothing strategies.

Shall I buy formula for "hungry babies"?
No.

Many parents are tempted by formula marketed for "hungry babies." These formulas give the impression that your baby will stay fuller for longer and perhaps sleep better. There is no evidence to support this.

Hungry baby formula is dominated by casein rather than whey, which might be slightly harder to digest. As there is so little evidence of any benefit, First Steps Nutrition Trust recommends that these formulas are only used under the guidance of a health professional.

Do anti-reflux milks work?

Anti-reflux milks are best avoided. We have very little evidence showing benefits of anti-reflux milks in otherwise healthy babies. Babies who need specialist milk will be prescribed it by the doctor.

Most babies bring up some milk at times, and this is called reflux. Both spit ups and occasional vomits are common in newborn babies and can continue for months, although they tend to start getting better after about three months as the gut matures. While regurgitation can feel alarming for parents, very few babies require any further investigation, treatment or a change from number one formula.

Anti-reflux milks are based on cow's milk but contain a thickener which is usually potato starch, carob or locust bean gum. These milks are marketed as foods for special medical purposes and should only be used under medical supervision. If you feel that your baby is struggling with reflux, please seek out specialist feeding support first to investigate exactly what is happening for your baby. Sometimes a change in teat, angle, pace, amount or frequency of feeds can make a significant difference. Most babies will grow out of it.

If your baby continues to show significant, frequent regurgitation with clear distress, your doctor might suggest that a thickener is given on a spoon or that you have a trial of thickened formula. There is some evidence that these thickeners can reduce the amount of regurgitation, but there is very limited evidence to support them as a "cure" for reflux.

Powdered anti-reflux milks appear to go lumpy when made with hot water, and the manufacturers suggest using water under 70°C. This is not in line with current recommendations for making up powdered infant formula safely.

Should I buy soy formula?

Soya-protein formula is not recommended for babies under six months.

Soya is a common cause of allergy in new babies. There are concerns that phytoestrogens in soya may impact sexual maturation. These formulas are also more likely to damage your baby's teeth. Although they are in supermarkets, please do not use these products unless medically recommended.

Is goat's milk formula better than cow's milk?

Cow's milk and goat's milk formula are similar.

Several brands of formula are made from goat's milk, and they claim to be easier to digest or kinder to baby's stomach. We have no evidence for this. Very rarely babies are allergic to the protein in either cow's or goat's milk, there is no difference.

How would I know if my baby was allergic to cow's or goat's milk?

It is unusual for a baby to have an allergy to cow's or goat's milk. If your baby shows signs of an immediate and severe reaction when they are given formula, call 999 for an ambulance. If symptoms are mild, call your doctor.

Symptoms of allergy could include:

- Skin reactions – such as an itchy rash or swelling of the lips, face and around the eyes.
- Digestive problems – such as stomach ache, vomiting, colic, diarrhoea or constipation, blood or mucus in the poo.
- Hay fever-like symptoms – such as a runny or blocked nose.
- Eczema that does not improve with treatment.

Reactions are typically seen when babies have formula for the first time. The allergy might be immediate, within minutes of drinking the milk, or delayed by hours or even days.

Very rarely, cow's or goat's milk can cause severe allergy symptoms that come on rapidly. This is a medical emergency. Call 999 for an ambulance if your baby has swelling in the mouth or throat, wheezing, cough, shortness of breath, and difficult, noisy breathing.

Some allergic babies will need to be prescribed an extensively hydrolysed formula. In rare and complex cases, a baby might be prescribed an amino-acid based formula.

When will I need Follow-on formula?

Follow-on formula labelled 2 or 3 is not required. It should never be given to babies under six months, and there is no evidence that parents need to switch to follow on formula after six months.

Follow on formula contains slightly more iron than number 1 formula, but babies will be eating appropriate foods after six months and will not benefit from the additional iron.

Children over one year can be given whole cow's or goat's milk but do not need toddler growing up milks. Some plant milks such as oat milk are also suitable. Please check First Steps Nutrition Trust online for the latest scientific information on this.

Shall I buy lactose-free formula?

No. Lactose intolerance is rare in babies, so only a very small number of babies will benefit from lactose-free infant formula given under medical supervision.

Brands including Aptamil and SMA offer a lactose free formula. Please do not self-diagnose lactose intolerance in your baby or buy lactose-free formula just in case. Parents who suffer with lactose intolerance are unlikely to have a newborn with lactose intolerance.

A baby struggling with lactose intolerance will be extremely unwell. If your baby is diagnosed as lactose intolerant, they will be prescribed appropriate formula by the doctor.

Some babies get temporary (secondary) lactose intolerance after a tummy bug, when their parents will notice green poos and discomfort for a few days. This resolves by itself and there is no evidence that lactose-free milk is beneficial in these cases.

Lactose free milks use glucose syrup as the source of carbohydrate. The glucose syrup can cause tooth decay, so parents need to be careful with brushing their baby's teeth. It could also cause changes in gut bacteria that impact on gut function and calcium absorption.

Can I make my own homemade formula?

Absolutely not. This could be very dangerous for your baby's health.

You might read recipes for homemade formula online. Please do not make your own formula. It is extremely difficult to get the right balance of constituents mixed appropriately, and there is a high risk of long term, irreversible harm to your baby.

Should I offer my baby water or juice in the bottle to keep them hydrated, or if they are constipated?

No. Formula (and breastmilk) already contains an appropriate amount of water to keep your baby hydrated, even if they are unwell or it is hot. Juice is not for babies.

Please offer your baby formula when they are thirsty. Water can be introduced in a sippy cup after your baby is six months old with their meals. You do not have to give water, they can have milk in their cup. This is for fun and learning, not as a primary source of nutrition.

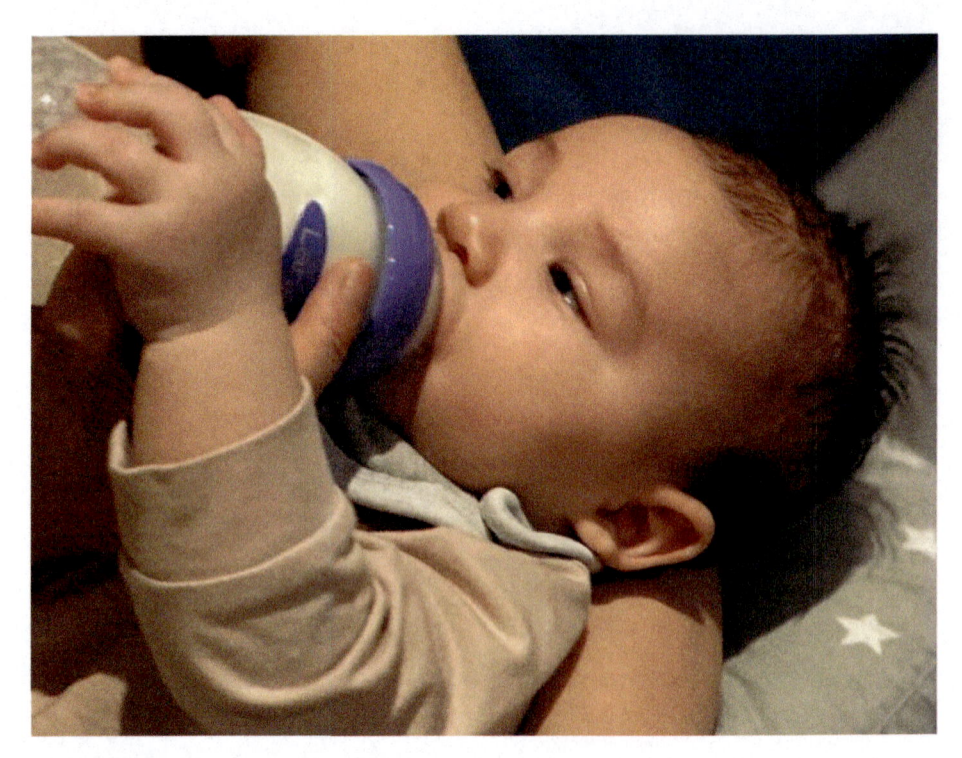

Are there infant milks for vegetarians?
Check the packaging carefully.

Not all manufacturers label formula milks as to their suitability for vegetarians. Specialised infant milks derived from cow's milk are generally not suitable for vegetarians due to the inclusion of fish oils and/or the use of the animal-derived enzyme rennet during the production process. Rennet, a by-product of animal processing, is used to separate curds from whey and, although vegetarian alternatives are available, manufacturers of infant milk do not typically use them. Some infant milks use constituents derived from pork or meat protein or use pork enzymes during the manufacturing process (even though this this may not be present in the final foodstuff).

Aptamil is releasing a plant and dairy based formula 3 for older babies, but this is not appropriate for newborns.

Is there specialised infant milk for vegans?
To date, there is no vegan formula.

Vegans avoid all products derived from animals and the vitamin D in infant milks comes from animal products.

Where a vegan mother requires a top up or replacement for her own breastmilk for one or more infants in a neonatal unit then donor human milk should be the first line of treatment. For more information on feeding vegan infants see the resource "Eating well for vegan infants and under-5's" at First Steps Nutrition Trust.

What about Halal and Kosher specialised infant milk?
Many specialised infant milks have sought approval for use by communities who require halal products. Many of those who choose a kosher diet will use specialised infant milks that are vegetarian or halal approved.

The term halal or "permissible" can be applied both to ingredients and to the manufacturing process. The approval process tends to give halal approval status to the factory or production facility to indicate that it has been inspected and approved to produce halal food, rather than to the product itself, so infant milks of the same brand which do not contain any forbidden ingredients may not all be halal approved (even if only the container size differs) if they have been made in different production facilities. It is important that this is checked for each milk if it is a concern to the family seeking advice. The First Steps Nutrition Trust asked manufacturers for the halal status of all products, but the information is not readily available. Parents must assume that a product is not halal approved unless specifically told it is.

Preparing your baby's bottles

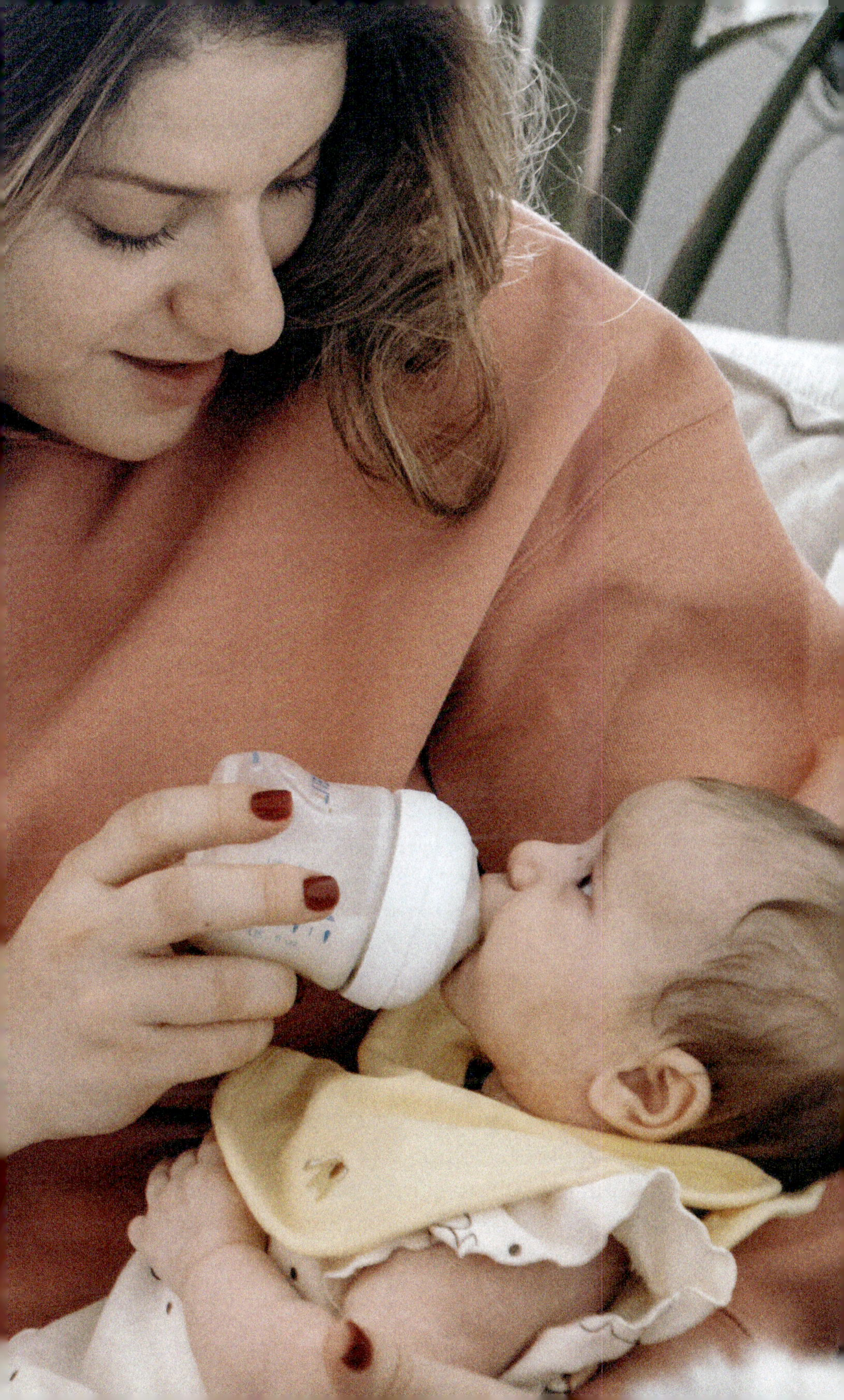

How do I wash and sterilise the equipment for my baby's bottle feeds?

It is vital to wash your baby's bottles and teats thoroughly in hot soapy water, rinse in clean cold water, and then sterilise them before each use. You can sterilise kit by boiling it in water, in a microwave or steam steriliser, using a cold water sterilising solution, or in a UV sterilising machine.

Obviously, you will want to protect your baby against infections including diarrhoea and vomiting. Formula milk does not have natural antibacterial properties so needs to be prepared safely and placed into sterilised containers.

The NHS currently recommends that everything that touches your baby's milk needs to be thoroughly washed and sterilised before use. You are meant to continue this until your baby is 12 months old and putting goodness knows what into their mouth!

Wash your bottle feeding kit well in hot soapy water before use, and as soon as possible after feeds.

You can use a dishwasher at 65°C, many machines have a baby bottle programme. Putting feeding equipment through the dishwasher cleans it but does not sterilise it. Make sure bottles, lids and teats are facing downwards. You may prefer to wash the teats separately by hand to make sure they are completely clean. Teats also tend to drop to the bottom of the dishwasher unless they are put into a container such as the cutlery compartment.

The advice on sterilising varies between countries.

There are several ways you can sterilise your baby's feeding equipment. The simplest is to **boil it in water in a saucepan on the hob for about ten minutes**. Make sure that the products stay under the water. However, this is not suitable for all kit and does tend to damage the equipment over time.

Most parents opt for a **plug in or microwave steam steriliser**. Follow the manufacturer's instructions, making sure that the open ends of the bottles face downwards so that steam can rise inside them.

Microwave sterilising bags can be useful for those with limited kitchen space or when travelling. They can be used up to 20 times.

UV sterilisers are becoming increasingly popular, and have the benefit of the equipment being dry afterwards. They take a bit longer than steaming methods, and the machines are often more expensive to buy. Some experts doubt that they sterilise as effectively as steam might do.

Cold water sterilising solution tends to be the least favoured option, as the items come out of the solution with a chemical taste and smell. Please follow the manufacturer's instructions. Items take around 30 minutes to sterilise items in the solution. As you add them, make sure there are no air bubbles trapped in the bottles or teats. Your steriliser should have a floating cover or a plunger to keep all the equipment under the solution. The solution itself needs to be changed every 24 hours.

When you remove items from the solution, shake off the excess water. It is often recommended to rinse the items with pre-boiled water from the kettle (not tap water).

After you've finished sterilising keep the sterilised items in the steriliser or a suitable clean container until they are needed. Each manufacturer will suggest how long the products remain suitably sterile before needing to be resterilised.

Obviously, apply sensible hand hygiene before touching the now clean bottles and teats. Some people use sterile tongs to pick up the items. If you are going to take them out of the steriliser, put the teats and lids on the bottles straight away to avoid future contamination. Assemble the bottles on a clean, disinfected surface or the upturned lid of the steriliser.

How should I make up powdered infant formula?

It is critically important to make up powdered formula correctly. Powdered formula is not a sterile product so must be made with tap water over 70ºC.

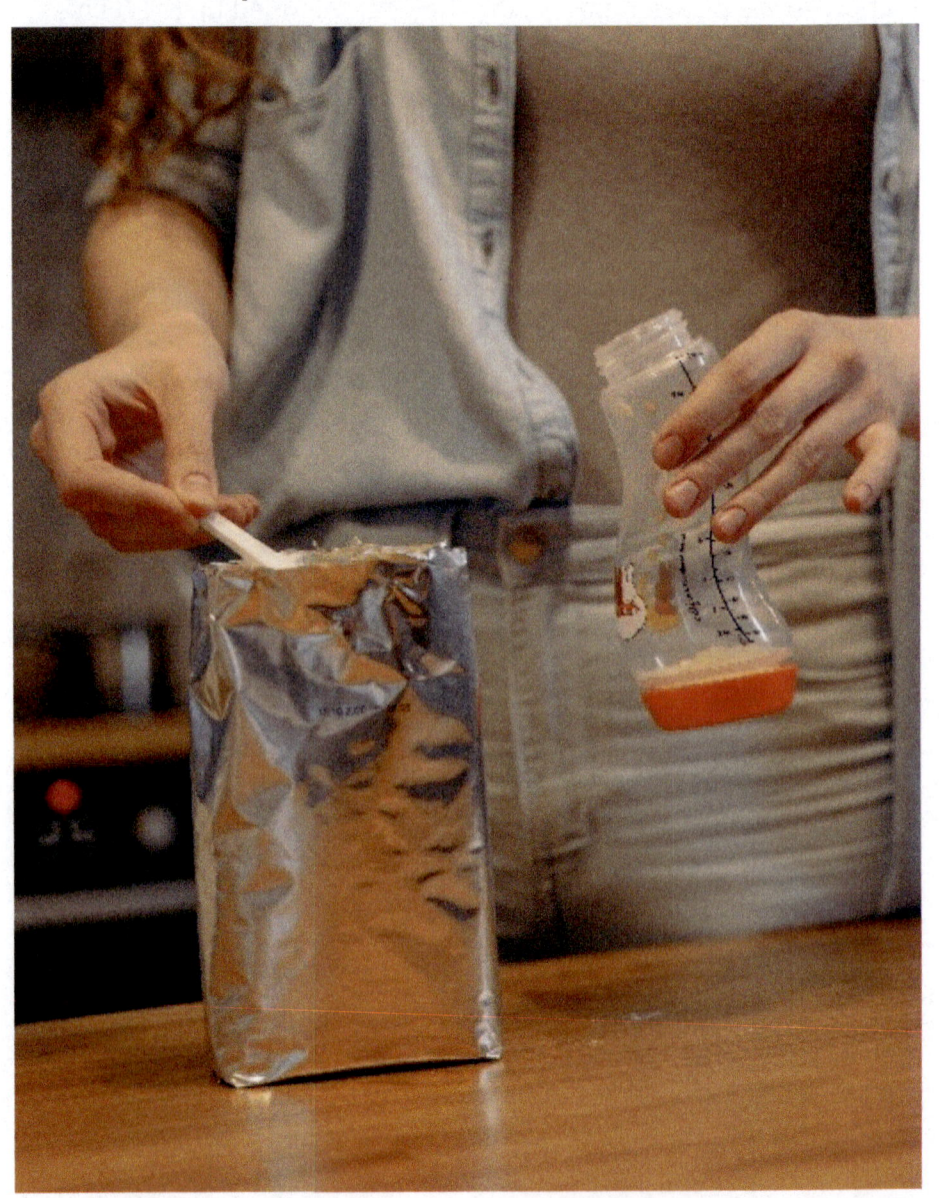

To make up powdered infant formula:

1. Fill your kettle with fresh tap water. Do not use bottled water, water that has been previously boiled, or softened water.
2. Boil the kettle, then leave it to cool so that it remains over 70°C. This will take up to half an hour.
3. Clean and disinfect the surface you will use.
4. Wash your hands with hot soapy water.
5. If you are using a cold water steriliser, shake off excess solution and rinse the products with previously boiled water from the kettle (not the tap).
6. Stand the bottle on the cleaned surface.
7. Refer to the manufacturer's instructions and make up the formula. This means adding the correct amount of water to the bottle first.
8. Use the scoop and leveller that came with that formula. Fill the scoop naturally without packing the formula down. Use the leveller or a clean dry knife to take away excess.
9. Push the bottle teat into the ring, pull it through, and then screw the ring onto the bottle top. Fit the clean lid of the bottle.
10. Shake or swirl the bottle until the powder is dissolved.
11. The formula is still too hot for your baby, it needs to be cooled to about 37°C. You can do this by holding the bottle (with its lid on) under cold running water. Bottle cooling machines are an alternative option.
12. The bottle is the right temperature when it feels warm or cool (not hot) if you drip a bit of formula onto the inside of your wrist.
13. Feed your baby. Left over formula must be thrown away after an hour or two.

Do **not** used bottled water to make up feeds. It is generally not sterile and may contain too much sodium or sulphate for your baby. If you are abroad and worried about the quality of the tap water, see the section How can I formula feed when travelling?

Do **not** add extra formula powder to the bottle. It can make your baby constipated or dehydrated.

Do **not** put too little formula powder into the bottle. Your baby will not get all the nourishment they need.

Do **not** add cereals, sugar or salt to your baby's milk.

Do **not** warm the milk in a microwave, it heats unevenly and could burn your baby's mouth. Do **not** warm milk in a saucepan on the stove.

Do **not** save left over milk in the bottle for the next feed.

What about bottle preparation machines?

Bottle preparation machines make feeds quickly, and many parents find them invaluable. However, they do not follow formula manufacturer guidelines and the resulting formula might not be as safe for your baby.

Many parents are using a bottle preparation machine to make up powdered infant formula more quickly than the standard approach. The machines put a shot of very hot water through the powder to kill bacteria and then adds cool filtered water to make up the bottle.

Recent research from the University of Swansea (2023) suggests that 85% of the 74 preparation machines tested in people's homes were not making the bottles hot enough to kill the bacteria that might be present in powdered formula. This is not following the formula manufacturer or NHS advice for making formula. In theory this could present a serious health risk, especially if the baby was small or vulnerable.

If using these machines, please take care to follow the instructions and keep your machine clean. You might like to use a thermometer to check how well your machine is performing.

Making up formula bottles one at a time sounds incredibly onerous, aren't there any alternatives?

Yes, but strictly speaking they are not recommended.

You may store prepared formula in the fridge for up to 24 hours, but obviously this is not as safe as using the product straight away before bacteria grow.

Many parents find ways to make up the formula bottle according to manufacturer's instructions but using slightly less hot water than required, and top it up afterwards using the missing amount of pre-boiled and cooled water to get a feed at the correct temperature.

This is not gold standard, as the concentration of the formula might not be the same as when you have prefilled the bottle with water and then added the powder.

You could consider using ready to feed formula from the supermarket, although this is more expensive than powdered formula.

Do I have to warm up bottles of formula?
Many babies prefer their milk at body temperature, which is about 37°C.

Ready made formula is kept at room temperature until opened, and the remainder kept in the fridge for 24 hours. If your baby likes their milk cool or cold, they will come to no harm. However, you can warm the milk in the baby's bottle by standing it in a bowl of hot water. Give it a swirl and pour some milk onto your inner your wrist to make sure it feels like a reasonable temperature.

There are several products on the market to warm and cool feeding bottles. It is not recommended to warm bottles in a microwave or on the stove in a saucepan.

How can I formula feed when travelling?
You have several options for making up the formula bottle away from home.

The simplest but most expensive way is to use ready to feed formula from the supermarket. Pour the product into a sterile bottle that you bring with you, or use a teat sold to fit directly onto the formula container. Unused formula in either bottle will need to be discarded after one hour.

If you are using powdered formula, you will need a flask of freshly boiled tap water. Ideally the flask should be just for baby's bottles, not also used for your soup or coffee. When you make the formula, this water will need to be above 70°C.

Travel with your clean and sterilised assembled bottles. Take pre-measured formula powder with you, in a clean container.

When ready to make the bottle, add the water first and then the powder, as you would at home, to make sure that the formula is the correct concentration.

If you have made formula in a sterile bottle in advance, it needs to be used within:

- 24 hours if kept in the fridge.
- Four hours if kept in a cool bag with ice packs or equivalent.
- Two hours if at room temperature.

The bacteria remaining in the formula bottle will still multiply if it is refrigerated, just at a slower rate than with a warm or room temperature bottle.

Take extra care if your baby is tiny or unwell. Fresh, ready to use formula might be the safest option in these cases.

Concerns about water quality.

In the UK, tap water is best, but if you're abroad and unsure of the water quality you may need to use bottled water.

Choose still (not sparkling) water, making sure that the seal is intact. It will still need to be boiled to make it sterile.

Mineral water might have higher concentrations of sulphate or sodium than recommended. Look at the label. You need a product where sodium (Na) is under 200 milligrams (mg) per litre, and sulphate (SO or SO4) is under 250 milligrams (mg) per litre.

Once opened, store the remaining bottled water in a fridge and use within the time recommended on the bottle (often three days).

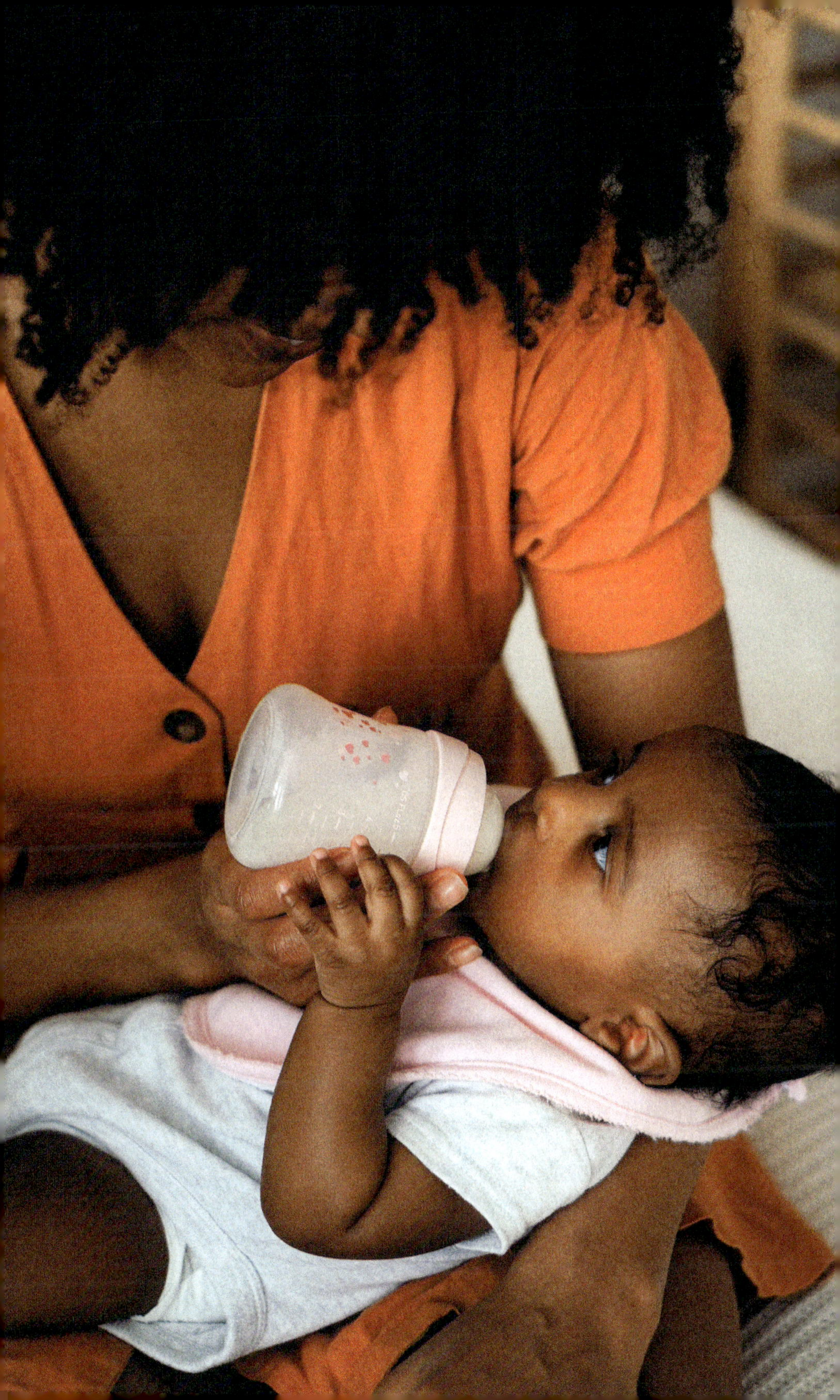

Your baby's first bottle feeds

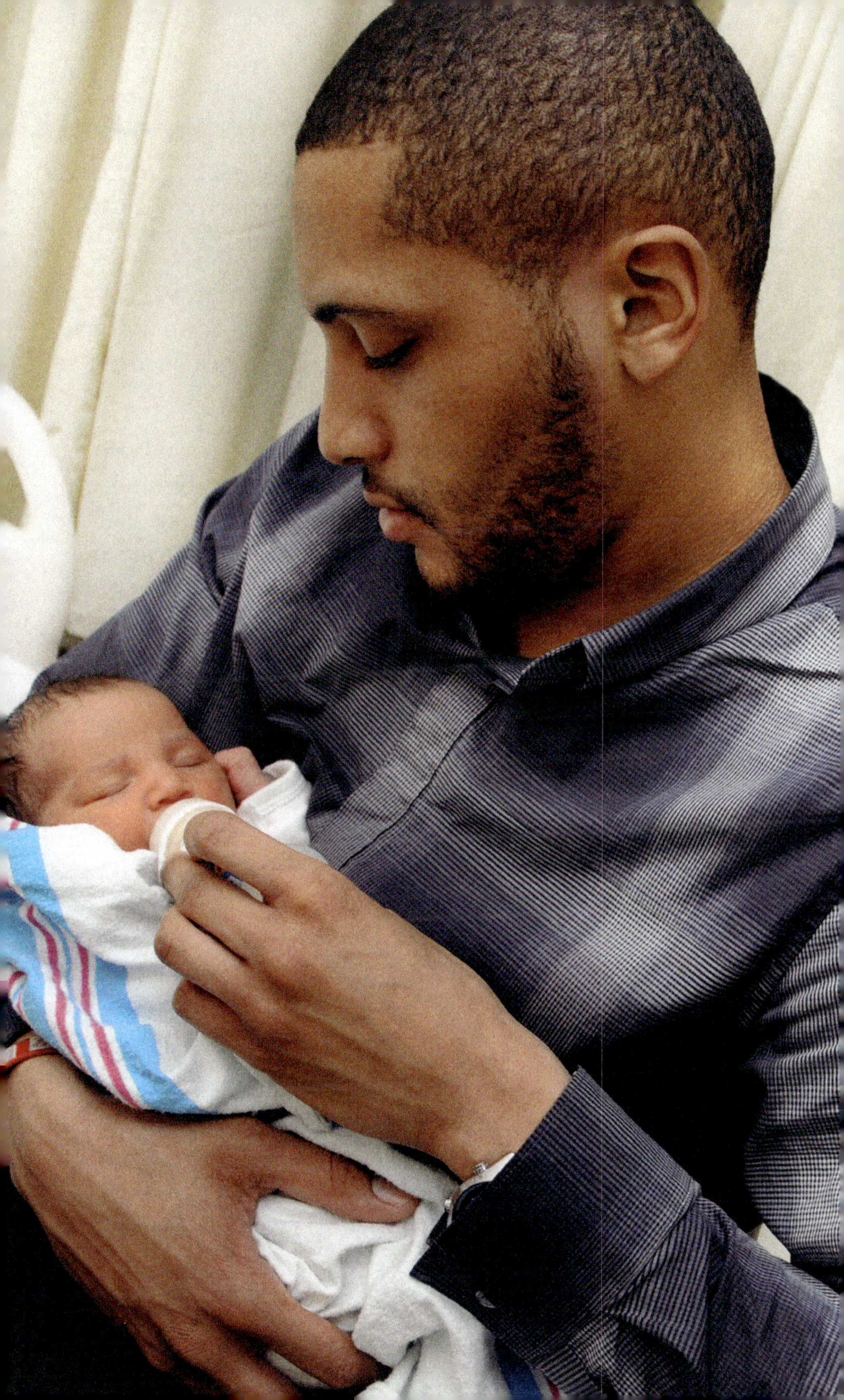

How do I start bottle feeding my newborn?

Ask for some help from your midwife. You might like to feed your newborn skin to skin, little and often, when they show feeding cues.

All babies love to be skin to skin with their parents. As soon as your baby is here, if you feel well enough, snuggle down with your baby on your chest skin to skin. There is no need to rush to get your baby weighed or dressed.

We know that skin to skin contact fills mothers and babies with wonderful happy hormones, reduces stress and promotes healing and bonding in them both. Babies respond well to hearing the heartbeat and voices they have been used to from those months in the womb. They do not need to waste precious energy in keeping warm, and they are colonising with their mother's friendly bacteria. It is lovely to offer your baby bottle feeds when they are skin to skin. You do not have to undress your baby to feed them every time, this is just a suggestion for the early days and when it feels practical.

How much shall I put in the first bottles?

Remember that your baby has never had a significant amount of milk in their tummy. They will need very small feeds, perhaps start with 10 millilitres (ml) on the first feed, and please watch them carefully to make sure that they can cope with the flow from whichever teat you have chosen, or that the hospital has provided. Some hospital bottle teats flow very fast for a newborn. The single use teats that are sold with formula also flow fast.

Breastfed babies would be taking a few millilitres of breastmilk per feed in the first days, but formula volumes will need to be higher than this. Be led by your baby's hunger cues. In the first few days it is reasonable to put 20-40 ml in the bottle and see how your baby gets on.

If your baby drinks their bottle, stops giving feeding cues, looks relaxed and maybe falls asleep, that was sufficient. If they are wide eyed, trying to suck their fingers and rooting around for more, then offer some additional formula. Look at the sections of this guide on nappy output and baby behaviour to be sure that your baby is getting sufficient milk overall and ask your midwife if you are worried.

Moving forward, follow the guidelines on the formula you have chosen, bearing in mind that many newborn babies take smaller, more frequent feeds to reach the same daily volume that the packet suggests.

Shall I warm the milk in the bottle?

Many babies, especially premature ones, would prefer their milk warmed to body temperature. We do this by standing the bottle of milk in a bowl of warm water.

How shall I hold my baby to feed them?

Bottle feeding should be as calm, happy and responsive as possible.

When bottle feeding, your baby will be most comfortable sitting up or slightly reclined, snuggled into you. They will want to have their head, neck and body facing in the same direction, rather than having to turn their head to feed. Their head will be held naturally or tilted slightly back, and they will appreciate some support for their back and neck. Many babies feel more secure if their feet are touching something.

Talk to your baby, stroke them, kiss their heads, just like you might when cuddling them.

Just for the record, when you see midwives feeding babies, they tend to hold them away from their bodies. This is because it might seem weird if they snuggle your baby. Your baby will probably prefer to be cuddled closer.

Some parents of twins must feed one or both babies in a baby chair out of necessity. Always watch your baby and do not prop up the bottles and walk away.

It is not good practice to bottle feed a newborn when they are lying flat on their back, as the pace of the feed is often too fast, and they cannot tell you to stop.

How shall I start the feed?

Use responsive and paced bottle feeding, as described below. Feed your baby when they are showing signs of hunger.

Tempt your little one to open their mouth by gently stroking their top lip with the bottle. Allow them to draw the bottle into their mouth. It needs to go quite deeply in, otherwise their reflexes might make them spit it out again.

Once your baby is sucking on the teat, keep the bottle tilted up just off horizontal, so that your baby can drink comfortably and not be overwhelmed with milk. There should be milk in the teat to the level of your baby's top lip. If the teat is too low there is a chance that your baby will suck in air not milk, if it is too high the flow can become overwhelming.

If your baby wants a break they may stop sucking. When you see this sit your baby up or tilt the bottle down so there is no flow. Start the feed again when your baby begins to suck. This makes bottle feeding much more like a breastfeed. Most babies need a little pause every so often to catch their breath. It also allows us to figure out whether they are still hungry, and we want to be led by their hunger rather than forcing them to feed.

If your baby seems distressed, stop the feed and perhaps offer them chance to burp. If they are repeatedly distressed, seek some help. You might need to change technique or equipment.

How will I know when to stop?

Your baby will gradually stop sucking much, will look sleepy, and might even push the bottle out of their mouth. Remove the bottle and watch your baby, to see if they show any further feeding cues. If not, they are full for now.

If they are sound asleep, you might choose to let them rest. If they are awake, you could try gently winding them for a few minutes.

End the feed when your baby wants to, even if there is still milk left in the bottle. A little bit of milk remaining in the bottle is perfect, it indicates that the size of that feed was just right for your baby. You will need to throw this milk away after an hour or so.

Please never force your baby to drink. This is disrespectful, unnecessary, and might result in your baby becoming averse to bottles.

How can I tell if my baby is OK with the bottle feed?

Watch your little one for signs of stress.

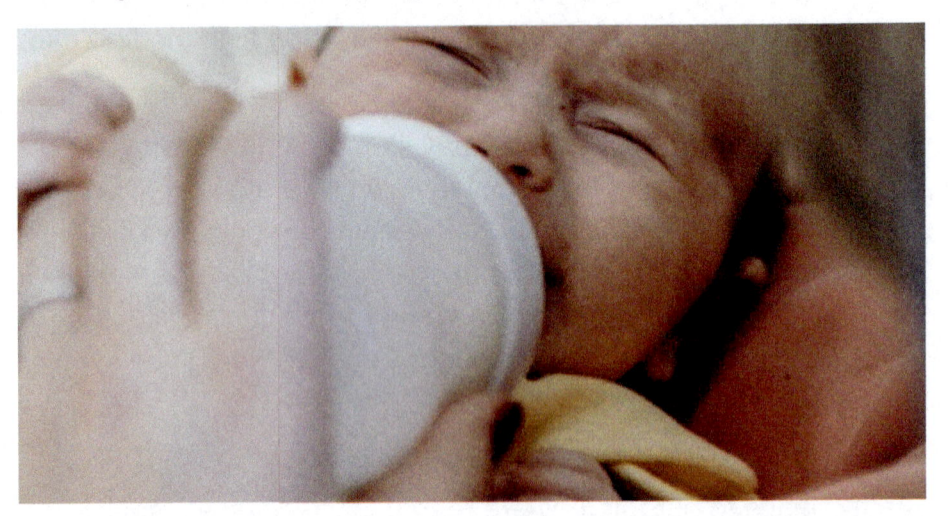

Bottle feeding should be a relaxed and enjoyable experience for you and your baby.

Your bottle feeding baby is showing signs of stress if they:

- Stop sucking and seem upset
- Dribble or vomit milk from their mouth
- Push out the teat
- Cough or gag
- Look red (in white skin tones)
- Arch back or kick
- Bat the bottle away with their hands
- Splay their fingers
- Make fists

If you see any of these things, please stop the feed and allow your baby chance to recover. Maybe offer them an opportunity to burp.

If your baby is consistently uncomfortable and stressed with the bottle, you will need to change the bottle or flow rate.

Babies that are having a hard time feeding from any bottle should be reviewed by a medical practitioner.

Do bottle fed babies need more winding?
Maybe. Responsive feeding may help to reduce wind.

We do not have much high quality published research about wind in babies. Some babies swallow a bit of air along with their milk, some babies make quite a bit of gas in their digestive system. If your baby is uncomfortable and perhaps a little red (in pale skin tones) during or after the feed, then try winding them.

You can reduce the amount of air in the bottle teat by keeping the level of the milk approximately at baby's top lip while they are actively drinking, rather than bringing the bottle up and down a lot so they swallow air and milk at the same time.

Your baby can digest milk more effectively when they are fed a little more slowly, and perhaps kept upright for a few minutes after a feed rather than immediately placed on their back.

Many baby books suggest a lot of vigorous winding of babies. To be honest, a lot of bouncing, patting and jogging the baby up and down is just as likely to cause discomfort as to get rid of it!

A baby that is sound asleep at the end of the feed is not showing signs of wind, so you do not have to disturb them to wind them.

If wind is a huge problem for your baby, and they seem constantly uncomfortable, please have your bottle feeding reviewed by someone who understands. It could be an issue with the bottle, teat or flow rate. There could be an anomaly in baby's mouth, such as tongue-tie. It could be the bottle feeding technique. Or it could be normal baby behaviour where they just seem to get super windy at around four to six weeks old and it goes away on its own. We do not have independent research to suggest that bottles designed to reduce wind, colic or reflux work. They might help, or the claims might not be justified. As so many healthy babies are windy it can be hard to tell.

There is very limited evidence that over the counter winding remedies are effective.

Do I have to keep my bottle fed baby upright after feeds?

Parents are often told not to put their baby on their back straight after feeds in case they vomit back the milk. We do not have firm evidence for this practice, but it might work for your little one.

Holding your baby upright against you after a feed feels like common sense. Most babies like to be carried this way and they seem to keep their milk down better if they are upright. Wearing your baby in a well-fitted sling does the same job. Opinions differ in how long keep your baby upright for. Fifteen minutes seems reasonable, but feels like a long time in the middle of the night!

It is quite common for a new baby to spit up milk while lying on their back, regardless of whether you winded them or held them upright. We would worry about this with a premature or unwell baby (or unconscious adult), but healthy babies seem to cope.

How much will my baby need in the bottle as they grow?

This varies with time and between babies. Try to be led by signs that your baby is hungry.

The formula container will give you a rough guide of how much formula is appropriate for different ages of babies. If your baby is well and weight gain is appropriate, it does not matter whether your baby is drinking a bit more or less than the chart recommends. Follow the guidance on paced or responsive feeding, looking out for signs that your baby is full or wants a little more.

It is wise to think about your baby's intake over 24 hours, rather than worrying about each feed. For example, in the morning your baby might be rooting around for more food after 60 ml and happily take another 30 ml if you offer it. At the next feed your baby might fall asleep after 50 ml. That is to be expected. Overall, if your baby seems well and satisfied, is doing appropriate numbers of wet nappies and is putting on weight, then you can relax.

For unwell or underweight babies on a feeding regime, the following calculation may be helpful. After the first few newborn days, your baby might drink roughly 150-200 ml multiplied by their weight in kilos in 24 hours. Divide the total by the number of feeds they are getting. For example, a 4 kg baby needs approximately 4 x 150 = 600 ml in 24 hours. If you were doing ten feeds, that is 60 ml per feed.

Research suggests that breastfed babies aged from one month to six months take 750-900 ml in 24 hours. Translating this into bottles, seven feeds might be 110-130 ml, eight might be 95-115 ml.

Formula manufacturers suggest that formula volumes should be increased with your baby's age, although this corresponds to larger feeds less frequently. Overall, the volume they suggest is higher than a breastfed baby might receive.

Your baby would most likely prefer slightly smaller, more frequent feeds, than those recommended. Please feed your baby responsively. Offer a bottle when they are hungry and watch for signs they are full. It is not necessary to try to stretch out the time between feeds or wake your healthy baby to feed them (unless told to by a doctor).

Many new parents read the side of the formula packet and see those quantities as a target. They fear that their little one will not grow unless they take all that milk. However, the formula company must state the highest amount that could be required, so that even the hungriest baby is properly fed.

Babies are instinctive, they know how much they need and will tell you both when they are hungry and when they do not want any more. Please respect that their body knows what it is doing. If your baby is happy and healthy in all other ways, and putting on appropriate amounts of weight, you do not have to stress about exactly how much they are eating on each occasion.

Some doctors suggest that bottles should be made of aluminium so that parents can't tell how much their baby has taken! There is some sense in that.

Can you overfeed a baby?

It is possible to overfeed if the flow is very fast, or you force your baby to finish their bottle even when they are full. It is also possible to consistently offer your baby more milk than they need in 24 hours. Your midwife or health visitor can help you to assess this.

The amount that your little one will want to drink depends on many factors, including the time of day, time of the last feed and what is going on for them developmentally. The side of the formula packet or bottle will give you a rough guide of how much your baby needs every day. When combining breast and bottle, you will never know how much your baby gets at the breast. Never mind, be led by your baby's behaviour. Feed your baby when you see feeding cues, stop when they want to stop. It is tempting to make them take the last drop, but that is not very fair. Your baby can't stop you from feeding them if the teat is in their mouth.

If you feed your baby very fast, sometimes their body does not pick up on the sensation of fullness as quickly as necessary. If your baby has far too much milk too quickly they are very likely to vomit it back up. Try slower or smaller feeds, a slower flow teat, and/or reduce the gap between feeds.

If your baby puts on weight much faster than is indicated, you may be overfeeding. Or maybe that is just your baby. Please discuss this with your GP or Health Visitor. To be completely honest, once your baby is crawling and walking these things tend to even out on their own.

Why is paced bottle feeding important?

Responsive, paced bottle feeding is baby led. It is respectful and therefore has less chance of causing bottle aversion. You are less likely to overfeed your baby or override their feeling of being full. It is more like breastfeeding so might help a baby who needs to do both.

As we have seen, paced or responsive bottle feeding means responding to your baby's hunger cues. It means feeding gently, relatively slowly, allowing your baby to pause and not forcing your little one to finish a bottle. In general, it means not applying a strict feeding regime in terms of amounts or timings (unless you have been told to by a healthcare professional).

All these considerations make bottle feeding more natural, like breastfeeding. Responsive bottle feeding is particularly useful if you would like your little one to take breast and bottle. There is some evidence that bottle feeding in this way reduces the chance of overfeeding, and allows your baby control over stopping when they are full. This could be useful for them later in their life.

This is in stark contrast to having your baby lie on their back, with the bottle upright, willing them to drink it as fast as possible so that you can get on with something else, however tempting that might be.

Common bottle feeding concerns

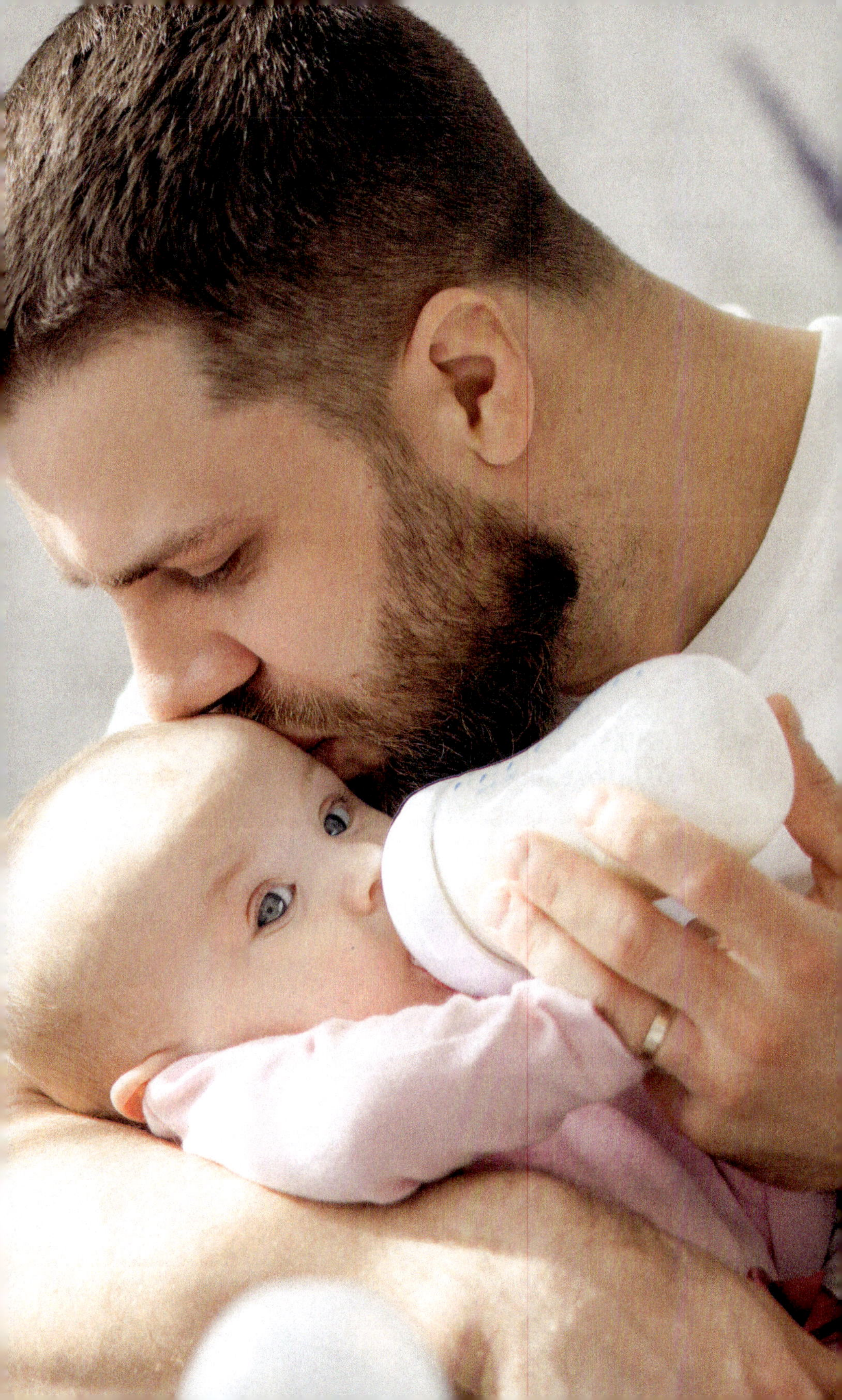

Is it true that only the parents should bottle feed the baby?

No. Do what works for your family.

Somewhere along the line, new parents have been told to limit the number of other people that feed their baby. The thinking goes that this might improve bonding between parents and the little one. Bottle feeding can certainly be part of forming a close and loving relationship with your baby.

However, taken too literally this advice has led to exhausted parents refusing help, and has left many excited grandparents, friends and older siblings feeling sad and left out! I have not found any sound evidence to support this practice, so do what feels right for you and your family.

If the mother bottle feeds will her baby get confused?

No. She might like or need a break, but many mothers bottle feed their babies due to time, necessity and logistics. Most babies take breast and bottle from the mother quite happily.

If have a reluctant bottle feeder, it might help mum to have a break while someone else attempts to feed the little one. On the other hand, some babies are much calmer if mum does the bottle feeding until everyone is used to it.

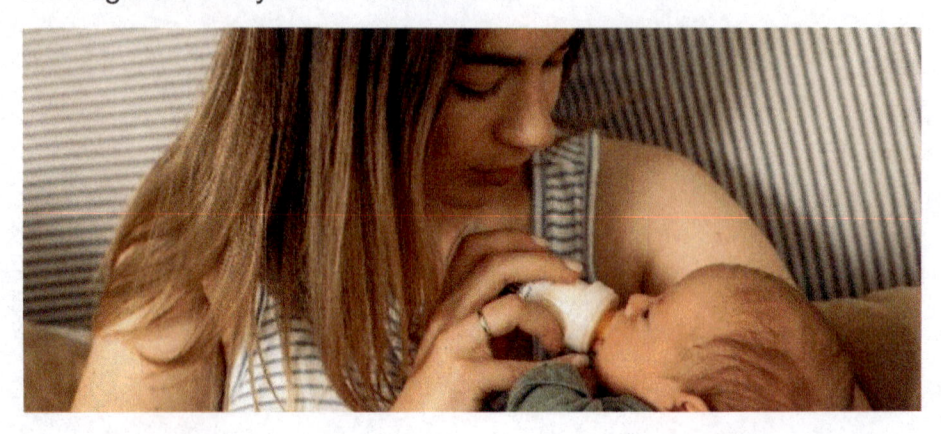

Will my breastfed baby reject the breast after having a bottle?

There is a lot of opinion about this, but precious little actual evidence. Many babies switch between breast and bottle seamlessly even from the early days. Occasionally babies refuse to go back to the breast after some bottles, and some breastfed babies refuse bottles entirely!

I often meet new parents who blame themselves for confusing their little one. This is unlikely to be the case. A baby who is having trouble with breast or bottle feeding is not confused, they are having difficulties.

Some parents are so worried about confusion that they refuse to feed their baby through a bottle even when their baby is dehydrated and desperate. This is not good for anyone.

Where does this leave new parents, who quite rightly want or need their baby to take a bottle sometimes?

First things first. **If your baby needs milk and cannot breastfeed please bottle feed your baby**. The priorities are to feed your baby, protect your milk supply by expressing, and get some breastfeeding support. If you are worried about using a bottle in the early days, midwives can help you to express or use formula in a cup or in a supplemental nursing system (a milk container with a tube that can be taped to the breast or a finger to feed the baby as they suck). Once your baby is taking a significant amount of milk, a bottle is usually the least stressful solution to use at home and is easier to keep clean.

It is a sensible idea to make bottle feeds as much like breastfeeds as possible. We discuss this in the section on paced or responsive bottle feeding.

If your baby resists breastfeeding after taking bottles, make sure that breastfeeding is going well and focus on it for a while.

How can I make bottle feeding more like breastfeeding?

Use the responsive, paced bottle feeding techniques that we discuss in this book.

Some manufacturers are selling products that are said to mimic the mother's unique breast size, shape and texture. Some partners choose to wear a bottle so it hangs like a breast. We do not have clear evidence that these products or practices are helpful.

If breastfeeding is going well, when should I try introducing a bottle?

There is no definitive research on when to start. A common sense approach is to wait until your milk supply is established and your baby is breastfeeding well, say four to six weeks old, before trying a bottle.

There is no rule that a baby ever needs to take a bottle, but many parents enjoy the flexibility of having a baby who will happily breast and bottle feed. People worry that any bottle feeding might upset breastfeeding, but we do not have strong evidence that this is the case.

Why wait for a month or so? The science suggests that breasts are still developing in the first few weeks after birth. They are adjusting the tissue to meet the demands of the baby, literally adding prolactin receptors in breasts that are used a lot and pruning down milk making tissue when it is not required. Leaving breasts full in these early days, for example, by missing night feeds, can result in less functional breast tissue overall.

Getting your baby used to a bottle before they are eight weeks may prevent bottle refusal in the older baby. We do not know this for sure, there is no proper evidence. You will probably need to continue offering a bottle at least every two to three days. If you are planning to return to work and need your baby to take a bottle, it makes sense to practice before the big day!

Any tips for giving the first bottle to a breastfed baby?

Use responsive feeding techniques. Here are some suggestions.

- Find a time when your baby is willing to eat, but not crying. This could be when waking from a nap, or when a little sleepy. Probably avoid late afternoon or evening when everyone is frazzled.
- Choose a teat size and shape that is appropriate for the age and mouth shape of your little one. You might like to have one or two different styles and flow rates to hand.
- Your baby will learn this new skill better if they are relaxed with someone that loves them. Mummy does not need to leave the building!
- Hold your baby securely, with lots of eye contact and reassurance. Skin to skin feeding is gorgeous.
- If you can, dip the bottle teat in expressed milk or formula first. Some babies prefer the milk in the bottle to be warm.
- Rub the teat on their top lip to ask them to open their mouth.
- Put the bottle into their open mouth facing up towards their palate so that they can make a seal and suck on the empty teat for a few seconds. Don't worry, they are not swallowing air.
- When they are sucking on the teat, raise the bottle so that they can access the milk. Keep the bottle teat full of milk, but bottle angled just off horizontal so that baby controls the flow of the milk.
- You want your baby to be swallowing happily. If they dribble, cough, gag, splutter, hit the bottle, pull away, flail around, make fists or splay their fingers, they are stressed. Stop for now and calm them. If this happens a few times, investigate a different bottle and teat. We don't try to bottle feed a baby that is crying.

- Watch your baby and give them a break if you feel they need it. Some babies like to be offered the chance to burp if they look red and squirmy, and after a feed if they are awake. There is no clear need to burp a baby that is happily sound asleep after their feed.

- If you are worried about your baby's bottle feeding, please reach out for skilled support. Some babies take time to learn this new skill, and some seem to forget how to do it when they used to bottle feed just fine!

- If your baby seems unwilling or unhappy about taking a bottle, please look at the sections on bottle refusal later in this guide.

If I mixed feed my baby what will happen to my milk supply?

It is a little complicated. Consistently reducing the amount taken out of the breasts will reduce milk supply.

Mixed feeding means different things to different people. It generally implies combining breastfeeding with bottle feeding, but the bottles could contain expressed breastmilk or formula milk or both through the day. How does that affect breastmilk supply?

This is a complex area. Breasts are primed to produce milk in response to demand from the baby. If you are expressing roughly at feed times and giving this milk at another time of day from a bottle, the breasts receive a similar amount of stimulation and demand as they would from your baby and supply should hopefully stay stable. This might be harder to achieve if you leave some very big gaps (several hours) between breastfeeding or expressing, because the breasts are left full and reduce supply. Every woman is different.

Breastmilk production is much more sensitive in the early weeks, and often more flexible and responsive after this. If protecting supply is important to you, and there are no other pressing reasons to give expressed milk or formula, holding off the bottles for about four weeks will give you the best chance of achieving your mixed feeding goals.

If you are planning to introduce formula feeds, it might be a good idea to do them at the same time each day, so that the breasts respond to what is required. For example, your baby might always get formula at around 10.30 pm. After a few days your body gets the message that breastmilk is no longer required in such quantities at 10.30 pm and adjusts to the new regime. This would happen naturally as your baby grows and drops feeds or stretches the time between feeds.

It is so hard to predict what will happen to each woman as they introduce more bottles of formula. Some women have flexible supply that adjusts perfectly. Others find that their supply goes down a lot, because their breasts need more frequent stimulation or emptying to keep sufficient production. Milk production can go up or down, so you can boost supply by increasing the breastfeeding or expressing again.

Bottle feeding issues

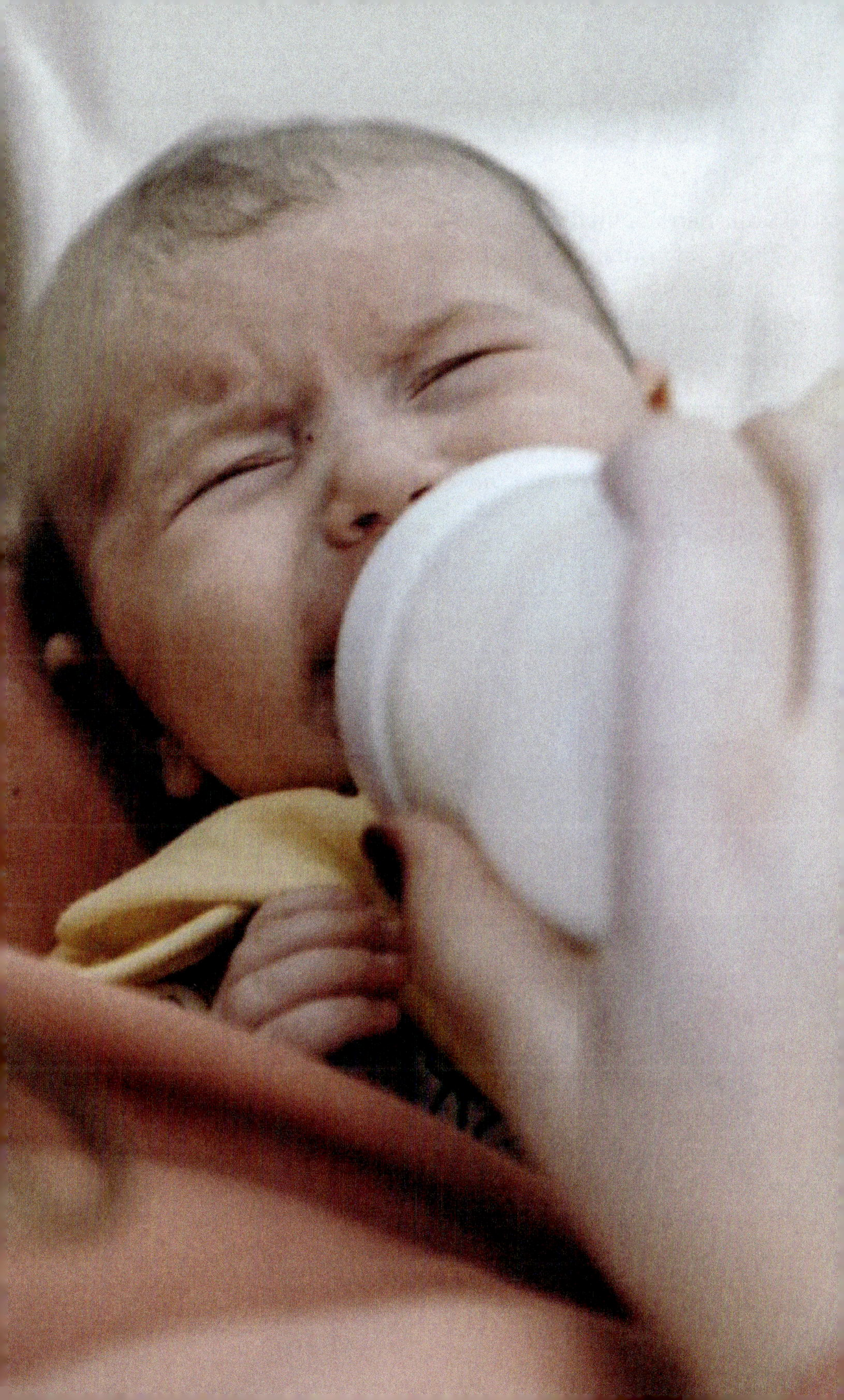

What can I do if my baby doesn't love bottle feeding?

If your baby is medically fine but still struggling with a bottle, please get some skilled support. We need to look at the big picture of how feeds are going and how to improve them. Changes in technique, tools and expectations might be necessary.

Some babies have problems with their mouth, tongue or airways that can make it incredibly challenging to bottle feed right from the start. Parents will be working closely with a team of medical providers to get their baby feeding in the best way possible.

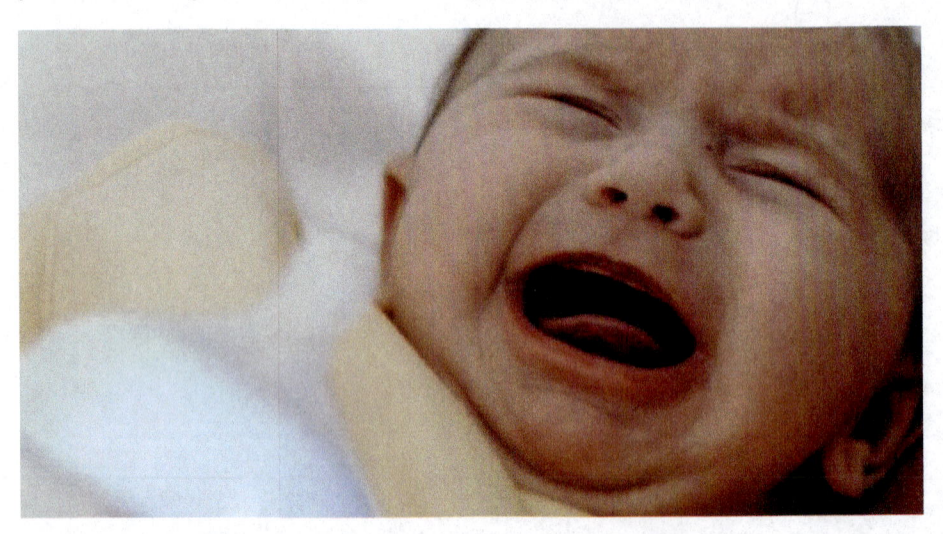

For babies without obvious issues, we will need a bit of detective work. You might be dealing with a baby that cries when they see a bottle or are held in a bottle feeding position. Perhaps they accept the teat in their mouth but can't latch onto it? Maybe they suck but then gag or choke? Maybe they are drinking too fast or slowly? Is milk leaking from the sides of their mouth when they feed? In each case, it might be necessary to review your bottle and teat selection and pick one that suits your baby. Then we can look at paced or responsive feeding techniques.

If your baby cries or arches back with what seems like pain, we need to review what is happening. It could be an issue with the speed of the feed, or it could be muscle pain, intestinal pain like wind or reflux. If you are using formula in the bottle, the cow's milk proteins and other components can occasionally cause digestive issues or even allergies in the baby. Allergies in babies are quite unusual, so please get your baby reviewed by a medical professional before self diagnosing or treating a suspected allergy.

When babies reject the bottle, parents often feel awful for doing something that makes their baby unhappy. Some will stop trying to bottle feed. Babies over four months old may be offered milk from a sippy cup or open cup instead. Alternatively, some parents decide to stop breastfeeding entirely and insist that their baby learns to take the bottle for all feeds.

What is bottle aversion?

Some professionals distinguish between "bottle aversion" when a baby gets distressed even at the sight of the bottle, and "bottle refusal" when the baby seems to be struggling with the mechanical process of taking milk from the bottle. There are no clear definitions, one may look very much like the other, and both could be happening!

We cannot be sure what is going on in the mind of a new baby. We know that babies who have a stressful or hard time bottle feeding often object strongly when they sense it is about to happen again. Personally, I avoid forcing a baby to feed in this situation, especially not pushing a bottle between their closed lips or persisting when they are crying.

However, parents at the end of their tether and some infant feeding midwives would disagree with me. They think that babies should be made to take the bottle regardless of their objection as it is ultimately for their own good. We do not have clear evidence either way. We discuss these approaches in more detail in the next section.

My newborn took a bottle and now they refuse it. What is going on?

Bottle refusal can be a huge challenge for parents and there is often no simple solution.

It is super common for a tiny baby to suck on a bottle OK, and their relieved parents then don't worry about it too much until wanting or needing to offer a bottle days or weeks later. At this point, many babies suddenly seem to have forgotten how to do it and can make their preferences very clear. Many of them squirm and arch back, hit the bottle, close their mouth or chew the bottle but not suck. Some babies happily play with the bottle but don't drink, others go crazy even at the sight of the bottle or when held in a bottle feeding position.

Parents often blame themselves, worrying that they didn't start bottle feeding early or regularly enough. We have no strong evidence that is the case.

Why do slightly older babies refuse bottles?

Babies are not being difficult, and they are not capable of manipulating you or "holding out until they get the breast." Developmentally, everything has changed. They have lost the initial reflex that made them suck on anything in their mouth. Their mouth and tongue have grown and the suck pads in their cheeks have reduced. It looks like they are pushing the bottle out of their mouth, but often they are trying to get their tongue around it to stabilise it before they can begin to seal their lips around it and suck.

Babies with a strong gag reflex might get distressed when the bottle touches their palate. Sometimes milk arrives before they are ready, and they gag or spit it out. Other babies don't suck and realise that the milk is there at all.

In my experience, the babies that have most problems with bottle feeding are used to a fast milk flow from their mum. They are typically not the first breastfed baby in the family. These babies commonly hold their tongue low and have a high palate. They have often been treated for tongue-tie, or have a suspected tie, but they are growing well on their breastfeeds. These babies have not been sucking very effectively at the breast, they have not needed to.

It is worth noting that around four months old many breastfeeding babies also have a change in behaviour, refusing to breastfeed during the day and catching up when they are sleepy and at night.

I need my baby to take a bottle, what can I do?

There are several possible strategies for a bottle rejecting baby. We have limited evidence around the best approach.

Before beginning to address bottle feeding issues, we need to be certain that your baby is well and growing within normal parameters. We need to be sure that your baby can take a bottle safely, so this means working with healthcare providers when necessary. If your baby is suffering from an anatomical or medical issue, these need to be well managed or resolved before working on bottle feeding.

Selecting bottle feeding equipment.

In many cases your baby is struggling with and objecting to the whole bottle feeding experience. Sometimes we can choose bottle teats and flow rates that suit their preference. It is often recommended to have a range of equipment to hand, to see which your baby prefers.

Babies that typically drink fast at the breast (feeding in under ten minutes) might prefer a faster flow bottle teat. Others might find this overwhelming and for these babies we select slower flow or variflow teat. We do not want your baby to cough or choke during a bottle feed.

If you are not sure about the flow of the teat you have, try sucking on it yourself.

While there is little independent research into bottles, we might find that smaller babies prefer narrow long teats, and almost all do better with longer rather than wider stubbier teats. Generally, a conical teat with a gentle slope, like the Lansinoh Natural Wave, is well tolerated. Some babies prefer the soft feel of a latex teat over harder silicone.

Offer your baby a bottle teat in its ring to suck on. Can your baby make a seal on it? Are their lips turned out and relaxed? If they can't do it, it is not the right teat for them, or they need a lot more practice.

Choosing milk.

It is great if you can express breastmilk for the bottle feed because we know that your baby likes it. However, wasting breastmilk left in a bottle can feel terrible, so number one formula milk of any brand is another option.

It is not recommended to buy lactose free, soy, anti-allergy or any of the other specialist milks available. Babies needing specialist milk will be prescribed it. Some prescription milks do not taste nice, so there might be a process of transition, for example mixing it with breastmilk, to get babies used to it.

Most bottle refusing babies like the milk warm, and it can help to invert the bottle to warm the teat or dip the teat into milk before starting the feed.

How much milk per feed?

Babies often resist the large volume formula feeds recommended by manufacturers. Think about intake over 24 hours rather than setting a target for each feed. Intake will vary from baby to baby and feed to feed. As a rough guide, a typical feed from one month to six months would be 90-150 ml. As with all things to do with babies, we read the baby not the book on this one.

A softly-softly approach with bottle refusers.

If you have the time, it can be least stressful for everyone to (re)introduce the bottle in a relaxed way. My personal preference is to keep bottle feeding light and fun, and to stop offering the bottle if the baby is very distressed. Nobody is at their best if they are wound up, especially parents and babies.

Your baby relies on you to help regulate their emotions, so it makes sense to keep the experience calm and light if possible.

If your baby becomes distressed at the sight of the bottle, it is worth getting your baby used to it in a non-threatening way.

You might keep several bottles around, allowing your baby to play with them. If your baby is old enough, you could hand the empty bottle to your baby to investigate. Alternatively, offer the empty teat in its ring like a pacifier. It can help to dip the clean bottle teat in expressed breastmilk or formula to make it super yummy for your baby.

Invite your baby to open their mouth to accept the teat. You can actively ask for their permission, "*would you like to try sucking on this?*" Obviously, your baby can't talk back to you, but it is surprising how all your interactions are a kind of conversation. Asking whether they want to feed is respectful. Forcing a teat into their mouth is not.

Once your baby is happily accepting the teat into their mouth you can start to play "tug of war" with your baby as they suck on it. If your baby will not take the bottle teat, you could practice this with a clean finger (nail side down), or with a pacifier.

When your baby is happy with a bottle teat in their mouth, it is time to offer a bottle with milk in it. You can read more about how to offer the bottle in the sections on responsive feeding.

If your breastfed baby cannot suck on a bottle, finger or pacifier after three days of trying, please get some support from a skilled professional.

Ideally, your baby will learn to take a bottle while you are sitting down, relaxed and comfortable. It might take a while to get to this point. If your baby cries in a bottle feeding position you might need to get creative! You might try bottle feeding somewhere novel like in the bath, bouncing gently on an exercise ball, or while walking around. You might vary the position, such as having your baby facing outwards from your lap.

Distraction can work wonders, babies sometimes start to drink while watching TV, being read a book or offered a toy.

Babies over six months might want to feed themselves, so bottles with handles for your baby can be great. Babies often love copying their parents, so be prepared to all sit around drinking from bottles of milk!

If all else fails, you might have most success with a bottle feed when your baby is super tired and has their guard down.

More robust approaches.

The remaining options might be required when your baby needs to have a bottle within a relatively short time frame. Two books detailing these approaches are by Rowena Bennett and Clare Byam Cook. **Before starting either of these approaches, please buy the books!** This can only be a high level summary to set you on a path that suits you and your little one.

Rowena Bennett wrote "Your Baby's Bottle Feeding Aversion." She believes that many babies become averse to being repeatedly forced or tricked into taking bottles. She asserts that we should respect our babies' preferences, and not offer the bottle more than twice in any given feed. Her theory is that when the pressure to bottle feed is removed, babies learn to trust their caregivers and take a bottle more willingly.

Rowena believes in respectfully offering the baby the bottle, asking their permission first. If the baby rejects the bottle in any way, by turning away, hitting it, refusing to latch onto it, or crying, the parent should simply remove the bottle. The baby should be offered the bottle twice in each feeding session. If they do not take it, the idea is to try again next feed.

Rowena believes that a baby will start accepting sufficient milk from the bottle once the pressure to feed is removed, and they are sufficiently hungry. If their baby does not take enough milk to keep them healthy, parents are encouraged to offer milk while their baby is sleepy at night, but only what is necessary to keep them hydrated. She does not offer a timetable for success, suggesting that it could take a few days to a couple of weeks.

The golden rules of Rowena's approach are:

- Baby should be under no pressure to feed.
- Baby should not be fed in their sleep during the day.
- Bottle feeds only.
- Follow baby's lead.

I must stress that this approach is only for babies that can take a bottle, therefore, it is not suitable for a baby struggling with anatomical difficulties or a medical condition.

Clare Byam Cook details a much stricter approach in her book "Top Tips for Bottle feeding." Her view is that babies are not traumatised by being forced to bottle feed, and that the sooner they learn to take the bottle the better. She asserts that the key to success is to get the baby to suck on the bottle by instinct not desire, and then make sure that baby gets the milk quickly and easily but doesn't choke on it. To achieve this, she suggests distracting the baby so they don't notice the teat going into their mouth, and hope that their response is to start sucking.

Clare suggests that it might take a full 24 hours before a baby accepts a bottle. She reassures parents that a healthy baby will come to no harm and will not harbour a grudge against their parents. The idea is to start with a hungry baby, so don't offer any other food for the previous four hours. During the next 24 hours, the baby should be offered nothing but the bottle – no solid food or breastfeeding. It helps to have more than one adult available in support, but generally the primary feeder should be the one to offer the bottle.

Once your baby has happily fed from a few bottles, Clare says that you can gradually reintroduce breastfeeding, but you should still give a bottle on a daily basis to prevent any further bottle rejection.

Clare Byam Cook's approach:

- Your baby is not going to remember this process, or hate you for it Warm the milk and use a relatively fast flow teat.
- Distract your baby, or walk around and pop the bottle into their mouth.
- Do not give up, even if your baby is unhappy.
- Offer nothing but a warm bottle of milk to your baby for 24 hours.

If you would like to try Clare's method, please do not start the process until you have read the full details about exactly how to do it as explained in her book "Top Tips for Bottle Feeding."

So, which is best?

Each of the approaches has advocates. We do not have decent evidence for what works best. It might be sensible to try calm and relaxed approaches first and keep the others available as a last resort. It is tough being a parent and this is one of the hardest things to deal with. Make sure that you look after yourselves through this process too.

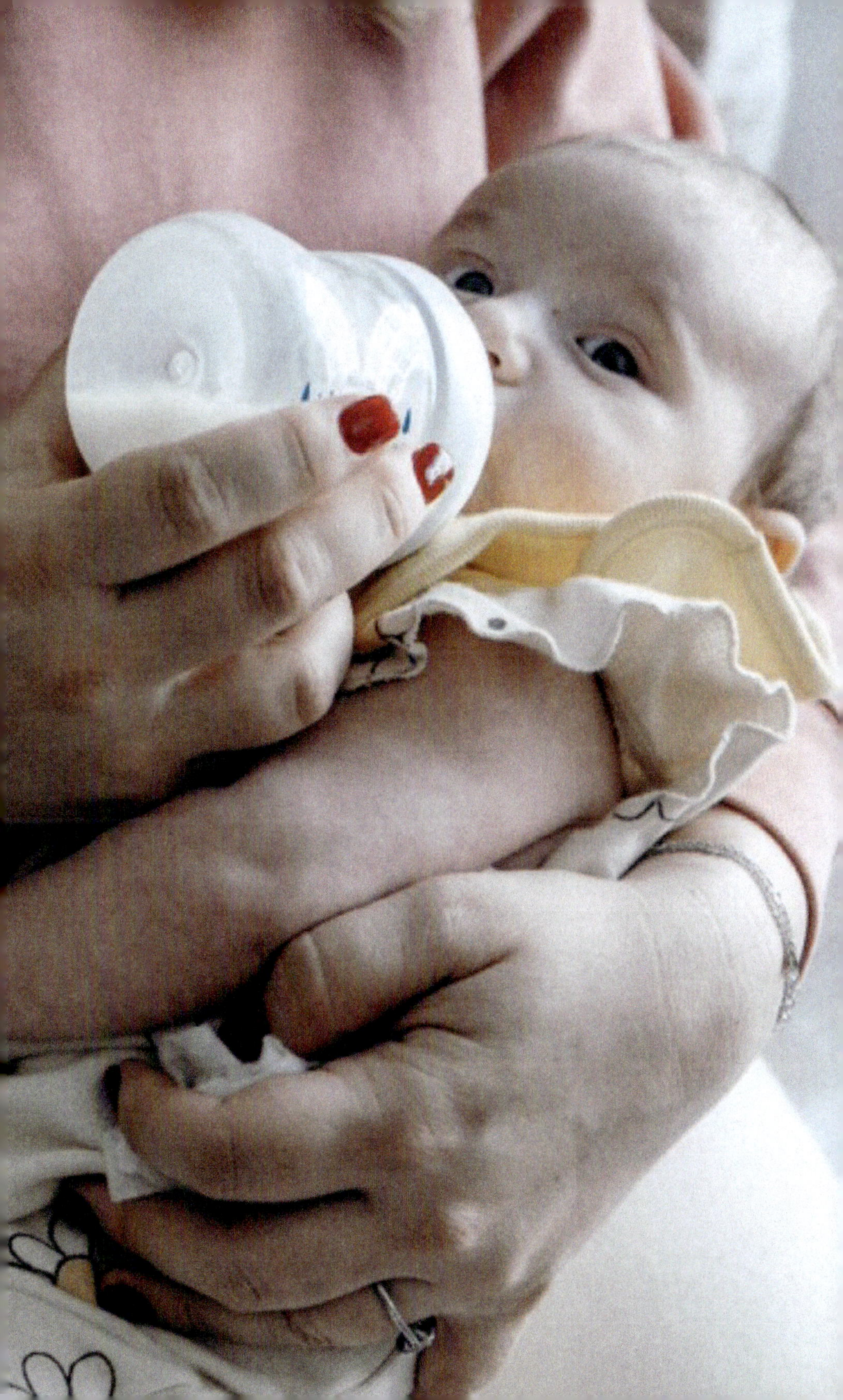

Baby feeding and your mental health

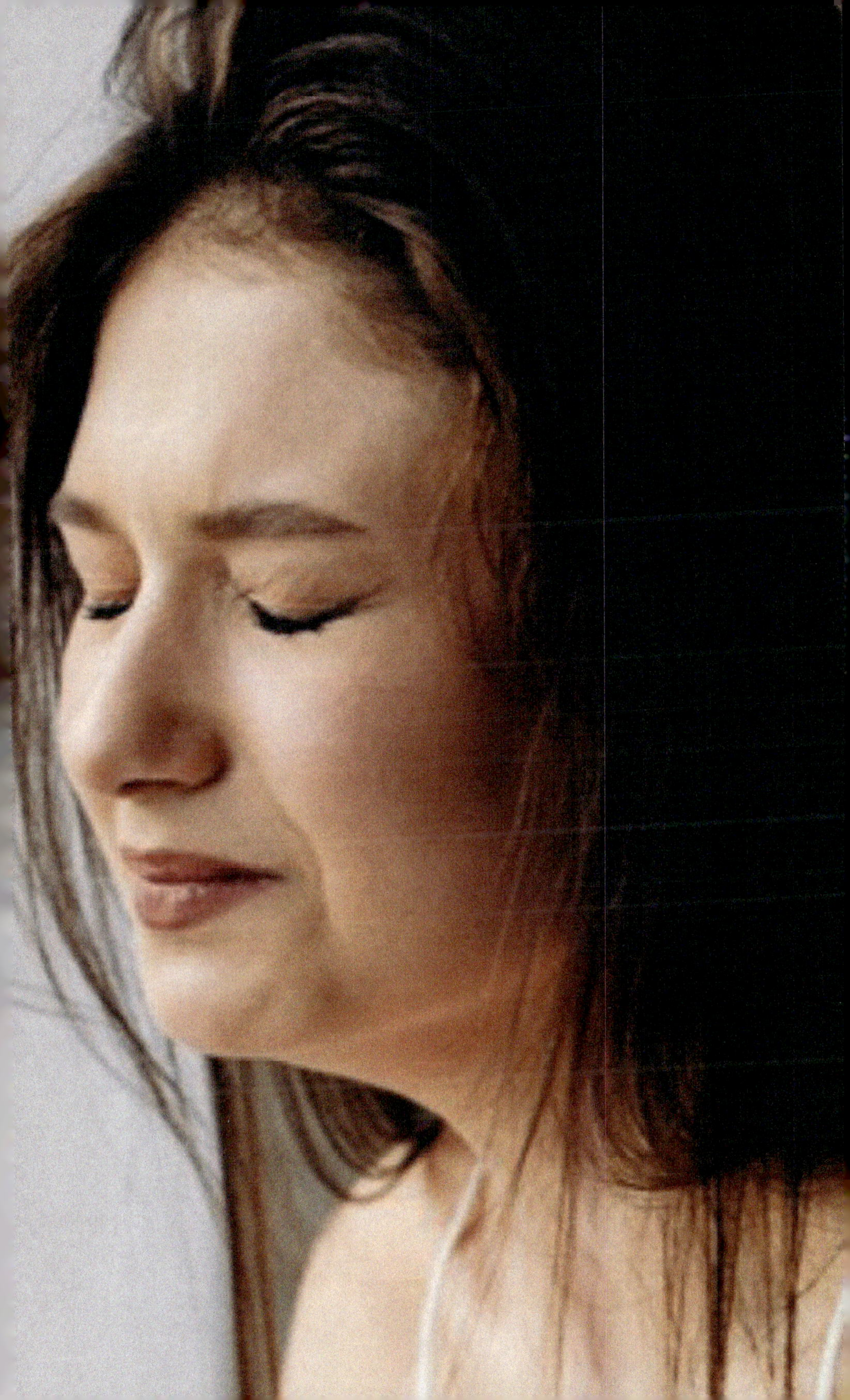

So far, we have talked about the physiological and practical aspects of feeding your baby. This is only a fraction of the whole picture. Having supported thousands of new mothers and their families, I would like to share more information about how you might be feeling emotionally alongside the breasts and bottles.

I hope that you will be happy and in love with your baby. Most new parents feel a complex mixture of emotions.

Nobody told me it would be this hard! Am I doing something wrong?

This is the first thing I hear from most of the new parents I meet. No matter how many classes they have been to, books they have read or items they have googled, nothing seems to prepare new parents for the reality of life with a newborn.

My own antenatal teacher, who I will always remember fondly, said that becoming a parent is like crossing a bridge to a new country. You can look back over at the life you just left but you have no way of getting back there. Everything is different now. Normal rules no longer apply. You can follow the manuals to the letter, but your baby is a little person and has other ideas. It might be the first time you have felt properly out of control. Not to mention exhausted. So deeply tired that you could sleep for weeks. Except you can't, obviously. Even if someone else holds the baby you are worried. You are constantly anxious that something might happen to the tiny helpless baby that you are responsible for, and it will be your fault.

We were not meant to parent on our own!

In the past, many of us lived with family who could help. Even if they were deeply annoying and opinionated, they could at least hold our baby while we cleaned our teeth, and perhaps feed us too. They could help us to know what was normal. Hold us while we sob and try to figure it all out.

The striking thing about life these days is that parents are trying to do it all on their own. Their own parents may live miles away and might be elderly. Many women become entirely isolated during the day as their partner returns to work. I appreciate that things are changing, but this remains the reality for most of us. Up to twenty percent of new parents suffer from postnatal depression. Yes, depression is a complex condition with many causes, but surely isolated and exhausted new parents trying to cope on their own would be at increased risk.

Who am I? I don't feel like myself

Becoming a new parent is a huge transition that should not be underestimated. The impact on your existing relationships with your partner, family and friends, is enormous and wide ranging. Most new parents are blown away by the changes and their implications, but as with so many things, these issues feel highly charged and difficult to discuss.

These feelings can be horrible. Nearly everyone has them, and very few people talk about them.

Becoming a new parent is incredibly challenging. It is very common for new parents to feel a profound sense of loss at a time when they expected to feel joyful.

I am also noticing an emphasis on gentle and responsive parenting that puts baby's needs above all else at all times, including the needs of the parents. This seems to be driving parents to **new heights of perfectionism and risk minimisation**, with a corresponding increase in shame and guilt. I don't think this is always helpful, parents must be OK too.

Many new parents are left thinking that they are freaks, and that they are the only people struggling to figure out who and what they are on the other side of the birth. Please rest assured that everyone is struggling. **It can take months or years to feel anything like "yourself"** and the new version of you will never be the same as the pre-child version.

Most new parents tell me that they

- Are overwhelmed and exhausted.
- Are lonely and isolated.
- Don't want to breastfeed all the time, but feel they must, or feel guilty/judged about using formula milk.
- Feel tied to their baby and not able to achieve anything.
- Miss work/colleagues and/or their previous life.
- Don't have any control or agency.
- Have lost a sense of who they are.
- No longer like or get on with their partner.
- Now have a changed or distant relationship with family or friends.
- Don't know the "right" way to do things.
- Are confused by other people's strongly held and contradictory opinions.
- Have little confidence in their parenting ability.
- Failed at birth, breastfeeding or parenting.
- Are anxious or low, or have scary intrusive thoughts.
- Are the only one not coping.
- Know they should be grateful for their baby, and confused about also feeing bad.
- Wish they didn't have a baby at all.
- Are worried that expressing any of this makes them a bad parent, or that someone will take their baby away.

The implications of parenthood stretch way beyond the physicality and practicality of having given birth to a baby (or babies) and needing to look after them. There is a word for becoming a mother, "matrescence," and there should be words for fatherhood and other partnerships too. As a society we need to give much more time, space and thought to honouring and supporting new parents through their transition. It is like adolescence or menopause or any of the massive life transitions, but bigger and harder.

New parents often tell me about their **profound sense of loss and grief** for a former self, lifestyle and place in society that they can never get back. This is rarely discussed but it is normal. It doesn't mean that they don't love their baby and the people around them, or that they are unfit parents. Like other sensible, thoughtful people, they are working through their transition and while it often becomes a little easier the process never stops.

Postnatal women are likely to struggle with a range of emotions and complicated feelings, including those around being **"touched out"**, being constantly interruptible, having no predictable time to themselves, always being responsible for a tiny being or beings that they love enormously but cannot always make happy.

The mother and new family can struggle with their baby's need to be fed so frequently and often unpredictably. Breastfeeding individuals can feel huge **resentment and exhaustion** as the "work" falls to them. Non breastfeeding partners can feel left out, or a little helpless, which is also difficult to cope with. Tempers can get frayed. Bottle feeding takes its own time and toll, but at least it can theoretically make the feeding aspect of parenting less one-sided and more equally shared.

Women may struggle with **body image**, especially when their body feels broken beyond measure. Scars, surgeries, stretch marks, pelvic floor issues, incontinence, piles, leaks from all sorts of places, the list goes on.

I suggest that all women seek out postnatal physiotherapy from a practitioner specialised in women's health. If you are facing any of these issues, please be kind to yourself and your beautiful body that has been through so much.

Very often, the flash points in a postnatal relationship are around **work and money.** For couples who have largely been equal up until maternity leave, the common transition to one not getting paid and the other providing, even if it is temporary, can be huge.

If you are at home, it is OK to miss work! It is normal to miss your colleagues, a sense of achievement, a salary, recognition, managing something, a job finished, a job well done. I even remember missing my commute. A chance to read on the train, listen to a podcast, the ability to grab a coffee and drink it hot. It is natural to miss this. For the partner at home, you now do not have that familiar life, or an income, or the chance to finish anything, or have a sense that you are doing something well. Life is an ever-repeating feeling of being exhausted, slightly (or completely) out of control and not good enough.

A partner returning from paid work can feel angry and resentful about coming home to a messy house, no food, and being asked or told to look after the baby. The pressure to keep a job can feel overwhelming. This applies whatever sex the working partner is. Both partners can feel enormously resentful, and then guilty about feeling that way. **Resentment is normal, but it hurts relationships**. For those in a romantic relationship, sex is unlikely to be on the cards. It takes a lot of openness, talking, patience and humour to maintain your relationship and it is wonderful to seek help if you can.

People keep telling me to trust my instincts, but I have never done this before! How can I trust myself?

I know, this is difficult and annoying. You will get more confident in time.

I wish that there was a manual that detailed exactly what your baby wants and needs and how to provide it. Of course there isn't, and you may be left worrying about each decision. In the end, by being with your baby a lot of the time you will soon become the expert on them, but it is still super hard to determine what the best thing is in any given situation.

Everyone gives us conflicting advice. Even the midwives contradict each other. What is going on?

I hear this every day. It is confusing and upsetting. There is huge variation in what is normal for each parent and baby.

Many health professionals including midwives, GPs and paediatricians are not extensively trained in baby feeding. On top of this, we just don't have the quality research studies on baby feeding we might like. An IBCLC qualification is gold standard in baby feeding, so you can reach out to one of us.

As part of my job, I am often around groups of parents talking to each other. I frequently hear them sharing information and references that are half true, or untrue, and sometimes positively unhelpful. It is even worse when outdated or inappropriate information comes from paediatricians or other healthcare providers.

Please take a breath and consider whether what you have been told fits with the big picture of what is going on for you and your baby. Does it sit well with your parental instincts? Is it based on evidence or opinion? Did it come from someone with something to sell? You might fact check what you have been told with evidence based information from the NHS, one of the national feeding organisations, or get a second opinion from your healthcare providers.

Sometimes I meet parents rigidly following advice that was relevant at a particular time, but nobody told them how or when to change their approach. For example, a feeding plan for an underweight jaundiced baby in their first few days can be adapted and relaxed as they get better. If you are not sure, please ask your midwife or health visitor, or seek out an IBCLC qualified practitioner for up to date information about baby feeding.

I hope that you will begin to learn about your baby, decide on your parenting style, and build a network of trusted supporters and professionals around you to sense check and guide (not tell you what to do)!

How can I cope with my guilt about my feeding decisions?

If you are feeling sad about your feeding decisions, you are not alone. Many new parents struggle with guilt about things that they have done, whether it was their choice or not. Reach out to like-minded friends and family, and perhaps even emotional support through therapy.

New parents often tell me about their difficult feelings, but rarely feel able to discuss them with peers. Perhaps your partially regret the breast reduction you had in your early 20s? Or perhaps you had a difficult start and your baby needed lots of formula, so you couldn't make enough breastmilk? Maybe you never wanted to breastfeed, but friends/family/society make you feel bad about it? You might want to breastfeed but know it upsets your partner or family who want to give your baby bottles. It is complicated and you might feel very conflicted.

It helps to have a sympathetic friend or group to talk to, find communities online, maybe get some counselling. Gabor Mate, a doctor specialising in trauma, asks his clients "When did you decide to be a bad parent? Was it last Tuesday? On the first day of your baby's life? Was it four years ago?" He aims to highlight that we are all trying our best, very few of us set out with bad intentions. You can't change the past. Right now, you just need to be good enough to keep your baby safe and loved. Feeling remorse about feeding decisions that were right at the time, or out of your control, does not help anyone.

Is it normal to not love my baby?
Yes. Sometimes it takes time to love your little one.

Just as there are many ways of falling in love, some people take time to get to know their baby and to even start to like them.

This is particularly common after a traumatic pregnancy or birth. Perhaps you had been scared that you were going to lose your baby, or your own life. It is totally understandable that some people feel ambivalent towards this little person who has caused so much havoc. We think that around half of brand new mothers feel a bit indifferent to their baby immediately after birth, but they don't talk about it.

If you feel this way, skin to skin contact with your baby, regardless of feeding method, can help. Most mothers feel better with time.

How can I deal with my feelings after a traumatic birth?

It can be very distressing to experience or witness a traumatic labour, birth or postnatal period. You may need skilled help ot treatment to promote your emotional recovery.

Many new parents find they need to talk about what happened, often repeatedly, as they begin to process it. I highly recommend finding someone who will listen intelligently and with empathy, with a focus on what you want for the future.

Most hospitals offer a debriefing service, where you can go and work through your birth notes, often with the team that was involved. This can help parents to get a sense of clarity and sometimes closure. However, sometimes it does not resolve the feelings that came along with the questions.

Most people find that memories of the trauma become easier to cope with after a few weeks, but some people find that it plays like a loop in their head, or can be triggered by certain situations, noises, smells or sounds. This is Post Traumatic Stress Disorder (PTSD), and it will be necessary to see a specialist.

There are some amazing services and therapies which can help a woman and her family begin to heal and move on from a traumatic birth. For example, the Rewind hypnosis technique has been proven to be effective in many cases. Another recognised form of treatment is Eye Movement Desensitisation. There are more details in the Sources of support section.

My family/partner are unsupportive of my feeding decisions. What can I do?
Listen respectfully, then try to do what is right for you!

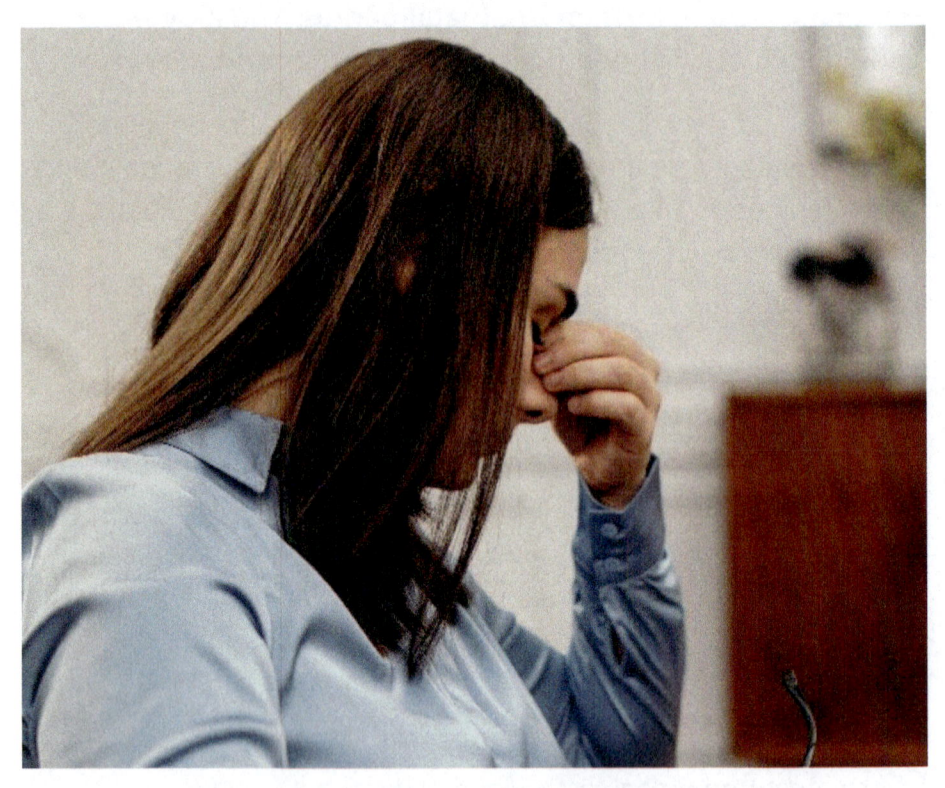

This scenario can be enormously challenging. Either subtle or direct pressure or judgment can feel quite overwhelming for new parents. Some people find their family extremely pro breastfeeding and judgmental about using formula even when it is completely necessary, in other cases it is the exact opposite.

Some people find it helpful to take a deep breath and listen respectfully to the other person's point of view. Ask some questions, thank them for their advice and for showing their concern. Then, once they feel heard, gently counter with the way that you want to do things and give the evidence for your own opinions.

For example, "*I know that you found some evening bottles essential when you fed James. You are worried that I am tired, and that Olivia is hungry. I appreciate that you want to help. Breastfeeding is important to me, and I would be so grateful if you would make me a sandwich so I can feed her on the sofa.*"

"I know that when Katy was a baby you were told to give her water in hot weather. Now they say that breastmilk or formula is better. It is crazy that advice changes so much, don't you think?"

"My lactation consultant says that my baby really needs some top ups of formula. I know it is not your preferred option, or mine, but we must keep him healthy."

There is more in the section on How can I deal with advice?

This is one of the most difficult issues to face. Some of you might be in deeply coercive or controlling relationships, or cultural situations where it is not possible to counter an elder's or husband's opinion. I appreciate that some religious groups are strict about feeding methods and duration. Please reach out to friends, peer support groups or professionals if you can safely do so.

Someone gave my baby a bottle. Should I be worried?

One bottle of formula is unlikely to adversely affect your milk supply or breastfeeding relationship.

However, giving bottles without your consent is disrespectful. Please challenge this behaviour, even if it was done with the best intentions.

Does my stress affect my baby?

Some stress is normal. Your baby may feel that you are stressed, and this is natural. Extreme or prolonged stress can impact you and your baby, and it is important to get skilled support.

Many new parents are stressed out. Then they feel stressed about feeling stressed. They worry that their baby will pick up on their stress and be damaged by it. Of course, this can make them feel like failures and more guilty and more stressed. So, what is the truth of the matter?

It is complicated.

Having a new baby is stressful.

Having a new baby without your family or friends or usual support network around you is super stressful. Doing so while facing illness, bereavement, job loss, financial insecurity, or any other of life's stressors, is even more challenging. Please do not worry about stressing your baby out when you are already going through things like this.

We know that anxiety and a certain amount of stress is completely normal and sometimes even healthy. Most stressful situations are horrible but unlikely to have negative long term consequences for your baby.

The kind of stress that might have more of an impact on a family is **long term, severe and toxic**. This might include clinical depression in either parent, or family members struggling with abuse, addiction or domestic violence. These types of stressors need urgent attention.

Is it true that a baby can "catch" your stress?
It depends on the age of your baby and the severity of the situation.

Babies are designed to pick up on your responses to what is happening around you, that is how they learn. It is thought that babies experience emotions such as sadness and anger from about three months old, and there is some evidence that babies from about this age can be upset by the sound of shouting even when they are asleep.

Even young babies will have some level of response to how you are behaving, your tone of voice, facial expression, your heart rate and temperature, possibly even your smell. They are sensitive to rough handling and big changes in their daily routine. Babies' ability to pick up on these things start very young and get better attuned with age.

More complex responses such as separation anxiety, are experienced a little later (from six months to three years, often peaking from 8 -12 months).

If a parent is continually stressed it can start to impact on how they care for their baby and their ability to tune into their baby's cues. There is a body of research about the impact of parental postnatal depression on their child. In these cases, there is a clear correlation between the length and severity of the depressive episode and its impact. The negative outcomes are mostly to do with the lack of normal loving interaction between the parent and child, and the ability of the affected parent to understand and respond appropriately to the child's needs.

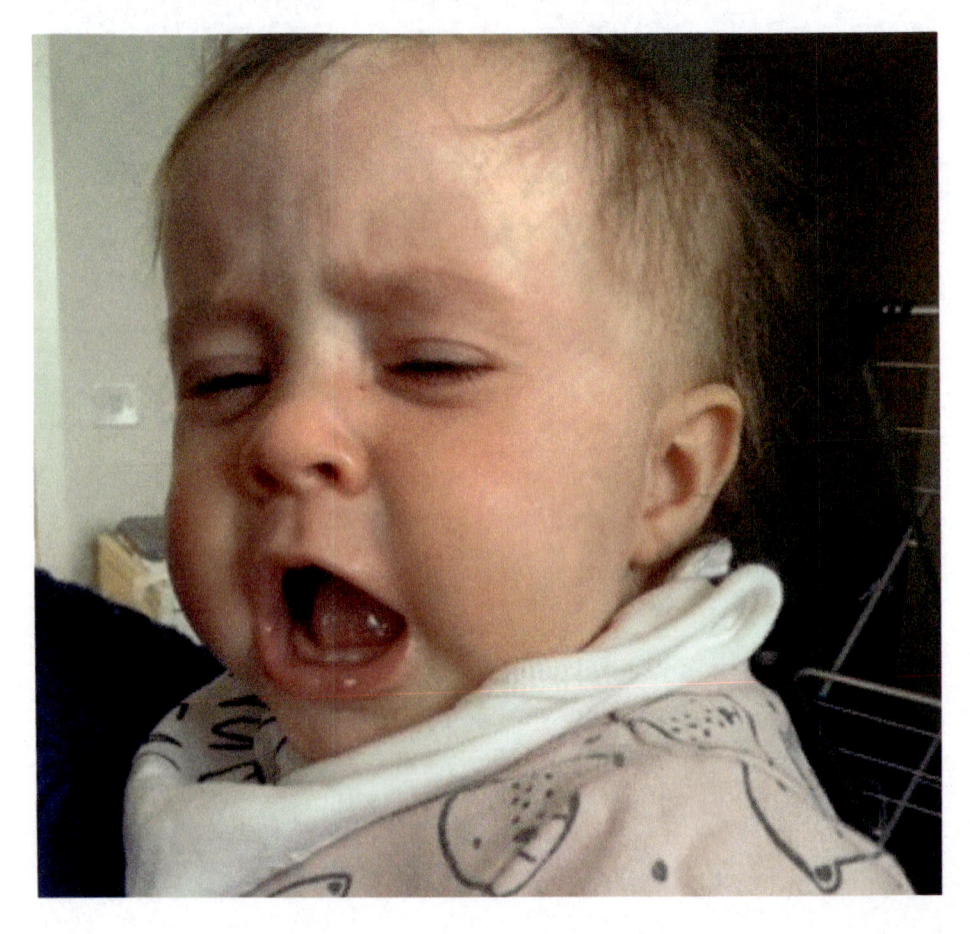

Are some babies "needier" than others?

Some babies seem to be born stressed out, and they are likely to result in stressed out parents, just as much as the other way around. Some babies have a massive need for stimulation and reassurance and seemingly constant close contact with their caregiver, and they might still be unhappy.

Please check that your baby is not unhappy due to illness, hunger or undiagnosed feeding problems. Skilled supporters can do a lot to help you overcome these issues.

Pamela Douglas' "*Discontented Little Baby Book*" discusses the "crying period" of the first sixteen weeks. Not all stressed out babies cry all the time. They might just get upset when you put them down. They might display other signs of stress such as avoiding eye contact, splaying their fingers, making fists, or yawning a great deal. They are unlikely to tolerate being put down in a cot for any length of time, and need to be held in arms.

Parents of newborns often see their friends' babies looking relaxed and happy, and wonder what they are doing wrong. It can feel alarming when your little one is upset and lead you to doubt yourself. It often feels like your baby will be like this forever. Many people tell me that their situation is "unsustainable."

Parents of needier babies need a lot of help and reassurance. It is a phase, but a totally exhausting and overwhelming one. Your baby is adjusting to the world and going through enormous changes. They are relying on you for their emotional and physical regulation, and often need to be held close a lot of the time.

You are definitely not spoiling your baby by holding them, and they do not need to learn to be independent. If you can, hold your baby a lot, wear them in a sling, enlist as much help as you can with the baby and everything else.

What can I do with a stressed out baby?

Obviously, make sure that your baby is OK. Address any potential feeding or health issues that might be making your baby unhappy. Then, try to remember that this is normal for some babies, that it is no reflection of your parenting ability or how much your baby loves you, and that things will get easier with time.

Parents have all sorts of ways of soothing babies, most of which will involve holding them close and rocking them, and perhaps shushing them. Wearing your baby in a well-chosen sling can make all the difference to everyone's stress levels. Babies love to suck for comfort, so generous breastfeeding can help a lot. If this is too much and baby just needs to suck, a pacifier can be a fabulous idea, or sucking on a caregiver's clean finger.

Many babies under three months respond well to being appropriately swaddled, although this is not recommended for nighttime sleeps and never when bedsharing with a parent.

How can I manage my stress?

Rest as much as you can. Accept offers of help. Ask for help! Fresh air, exercise and a nutritious diet all support wellbeing. Remember that this is understandably hard and to try to be kind to yourself.

If you are overwhelmed by breastfeeding, you might ask someone else to offer a bottle of expressed milk or formula so you can rest. This might be an occasional bottle or a regular part of your day. Please read the section on mixed feeding for the pros and cons of this. Giving just one bottle feed a day can allow an exhausted parent five hours of sleep in a row, which is found to be super helpful in maintaining mental health. Studies show that some parents can continue breastfeeding for longer overall if they get some rest in this way.

Reach out and connect with like-minded people. If you have the emotional strength to pick up the phone and call a family member, friend or another new mum, you are likely to feel better. If not, try to contact a professional. There are lots of support groups out there, on the phone or online. Sharing your experiences with other new parents allows them to be honest too, which is good for everyone.

Make friends that share your values. Try to build and maintain relationships with people who support you. If anyone makes you feel rotten about yourself, they are not ideal people to hang around with physically or online.

Be the first one to **be honest** to members of your support group or antenatal class. It gives everyone else permission to be real too. It was probably two or three months after I had my baby when the first woman was brave enough to say "*actually this is shitty and I am on anti-depressants.*" Then the floodgates opened, everyone told the truth and the genuine picture emerged. We became and remain much better friends.

Remember that how you feed your baby is nobody else's business. Babies need to be fed and loved. Feeling guilt or shame about feeding decisions does not help anyone.

Fresh air and exercise are a good thing. Try to leave your home each day if you can, even a quick evening walk can be helpful. Exercising in company is even better. If you can access postnatal yoga or pilates groups, they will help to build up endorphins and give you a sense of agency. If your baby is grouchy or it is officially nap time, try putting your baby in a sling or pram and do it anyway.

You might like to **get informed** about how others are feeling and coping. One of my favourite books is "*The Little Book of Self-Care for New Mums*" by Becky Hands and Alexis Stickland. There are some wonderful Instagram feeds from women who know what it is like, you might like to try posts tagged with **#fourthtrimester**, **#matresence**.

Try to eat well. Your diet does not need to be perfect. Some fresh fruit and vegetables would be good.

Readjust your expectations about what is normal. Prioritise looking after yourselves as much as your little ones.

Please do not expect much from yourself. Things will get better in time.

Try the mental health apps and websites. There is a full list in the Sources of Support section at the back of this book.

Remember that much of life's stress is unavoidable, it is how we deal with it that matters. Most normal life stress that we experience will not impact on our little ones in any meaningful way.

Is it normal to feel low in the early days? I am crying and I don't even know why.

It is normal to experience an emotional roller coaster in the first few days, weeks and months of being a new parent. It is generally not postnatal depression if a woman is sometimes overwhelmed and tearful, but other times happy and coping with her baby. If you are worried, please see your GP.

You might have heard of the **baby blues** that eight out of ten mums will get from about day four. At this time a woman's hormones are going through huge changes, her milk may come in, and her breasts might be swollen. It is a common time to feel overwhelmed and tearful. Baby blues can persist for a couple of weeks.

Even after the baby blues it is normal to cry about something every day. In fact, it would be bizarre if you did not have a hard time physically and emotionally adjusting to a whole new life with a baby. That is not postnatal depression unless there is never any lightness or joy to the day.

Do I have postnatal depression?

Postnatal depression is different from the baby blues. Parents' normal activities including eating, sleeping and social interaction will most likely be affected. If you are worried about depression, please see your GP. Proper support, Cognitive Behavioural Therapy (CBT) and sometimes antidepressants can be helpful.

Postnatal depression (PND) is extremely common after the birth of a baby. It is estimated that about 20% of women and 10% of men will experience it. Even in our new climate of talking more about mental health, it is certain that a huge number of new parents do not seek support from professionals or even friends and family. As depressed parents find it hard to leave the house, they can become increasingly isolated.

PND can manifest in many ways. Some people talk about a grey fog around them, inability to find joy in their lives and their babies, their struggle to get through the day. Many of them sleep a lot, many can't sleep. Some will eat excessively, some cannot eat. They might suffer with anxious and obsessive thoughts.

One partner told me:

"I had forgotten my key. As I rang the doorbell, I could see through the door to my wife sitting in the kitchen. As she heard the bell, it must have been the final straw. Her shoulders slumped and I saw her hit her head against the wall three times. That is when I knew this was much worse than new parent baby blues."

This is an incredibly under-researched area. Talking therapies such as Cognitive Behavioural Therapy (CBT) can help and are believed to be as effective as antidepressants in treating many forms of depression.

It is a great idea to get out of the house for fresh air, exercise and company, but this can feel almost impossible for a depressed person.

"We had to go back to the GP several times. Looking back, we were way past the point where the standard advice of "get some exercise and join a baby group" were going to be enough. My wife went on antidepressants, and they helped to stabilise her mood."

Your GP might suggest medications. Some antidepressant medications such as Sertraline are considered compatible with breastfeeding, discuss this with the doctor. Some women with depression might find breastfeeding therapeutic, others might find it unbearable.

There are some apps available that can help you to monitor your mood and celebrate small victories.

If PND is an issue for you or your partner do not ignore it, please go to the GP and start finding some treatment to help you. As depression and feeding are likely to be linked, please find local sources of support with this and other aspects of parenting.

What is postnatal psychosis? Can I still breastfeed?

Psychosis can come on rapidly after birth. It is a medical emergency. If a mother is hallucinating, delusional, manic, or thinks she might harm herself or her baby, she needs to be taken to hospital immediately. Although she is unlikely to be separated from her baby, breastfeeding might not be possible if either the mother or baby is in danger, or the mother is taking certain medications.

Very sadly, about one in 1,000 mums will suffer from a form of postnatal psychosis. This is a medical emergency, and the mum needs urgent treatment at hospital. It is nobody's fault, and it will get better with medical support. Nobody knows exactly why it happens, but mothers are considered more at risk if they are already diagnosed as schizophrenic or bipolar or have a family member who suffered with psychosis.

Symptoms of postnatal psychosis include:

- **Hallucinations**: seeing, hearing, feeling or smelling things that are not there.
- **Delusions**: believing, fearing or suspecting things that are not true e.g. my baby is the Devil.
- **Mania**: acting speeded up, talking or thinking too much or too quickly, restlessness, losing normal inhibitions.
- **Low mood**: showing signs of depression and low mood, having trouble eating or sleeping, overwhelming anxiety or agitation.
- **Switching rapidly** between high and low moods.
- **Feeling very confused**.

A mother suffering from psychosis is unlikely to recognise it. She may be resistant to getting treatment. It is often a partner, friend or family member who will need to call 999 or take them to A&E. It is extremely frightening for the woman; and partners, family and friends will also benefit from some support.

Ideally, a woman with psychosis will be admitted to a secure mother and baby unit where they can be properly observed and looked after. It is very unlikely that anyone will take her baby away.

It is possible that a mother will be admitted to a general psychiatric ward while waiting for a place in the mother and baby unit. It can be more difficult for a mother to keep her baby with her in this situation. Please discuss this with the caregivers at the time.

Various treatments may be offered, including one or more of the following.

Possible treatments for postnatal psychosis:

- Antipsychotics to help with mania, hallucination and delusions.
- Mood stabilisers to prevent the symptoms from coming back.
- Antidepressants might be used alongside the mood stabilisers.

Some mothers with psychosis within a specialised unit will be able to continue breastfeeding their baby, assuming they have sensitive supervision. However, some antipsychotic and mood stabilising medications such as lithium could make this inadvisable. It is vital to work with the medical professionals at the facility who are looking out for the best interests of both mother and baby.

Most people with postpartum psychosis make a complete recovery when they have the right treatment. The worst symptoms tend to pass in two to twelve weeks, but the whole episode can last for six to twelve months.

Action on Postpartum Psychosis has some excellent resources and sources of support.

Why am I having horrible thoughts about harming my baby?

It is common to have distressing thoughts about harming your baby accidentally or on purpose. Most people can recognise and dismiss the thoughts. If you are worried that you want to act on them, please get specialist medical support from A&E straight away. If you do not want to act on your thoughts but they are changing your behaviour, there are lots of ways to get help.

New parents worry a lot of the time, this is a brilliant strategy for anticipating danger and keeping their little one safe. A side effect of this can be nasty thoughts about hurting the baby. We think that nearly all new parents have upsetting thoughts. They are considered normal if you do not want to act on them and can stop thinking about them quickly.

However, sometimes the thoughts become a problem. The more you try to push them away, the more you think about them, and you begin to change your behaviour to cope with them.

When the thoughts are overwhelming, some parents become obsessive in the way they look after their baby. They might check their baby or the surroundings many times, perhaps continually taking their baby's temperature or compulsively weighing them. Some repeatedly seek medical opinions about their baby's health, but don't believe the doctors. Others disassociate and believe that their baby would better be looked after by someone else.

In these cases, please reach out for specialist support. The GP is a good place to start. I must stress that nobody wants to take your baby away, and you are not a bad parent for thinking these things. We want to help you to get new strategies to overcome the thoughts. The excellent book *"Dropping the baby and other scary thoughts"* by Karen Kleiman et al is a great resource. It is calming and reassuring, and offers strategies to work through. If you discuss these things with trusted friends, your GP or a counsellor, you might be surprised by how many other people are struggling in silence.

I am constantly anxious about my baby's feeding, what can I do?

Seek out some skilled support from your midwife or health visitor, a lactation consultant, feeding group, or one of the national or local parenting organisations. If you are told that your baby is feeding OK, but you are still overwhelmingly worried and changing your behaviour to cope, you might benefit from more skilled psychological support.

This is a difficult thing to write about, as it is totally normal for a new parent to worry about how their baby is feeding. However, I sometimes meet parents whose babies are fine, and they have been reassured by professionals that nothing is wrong, but they still struggle with obsessive or compulsive behaviours around feeding. I have worked with mothers who test weigh their baby before and after all feeds, or who insist that their baby feeds every moment they are awake in case they "drop a centile," or break down with anxiety when their baby does a normal posset or vomit. This is a challenging area. There are skilled professionals who can help, and it is great to reach out for support.

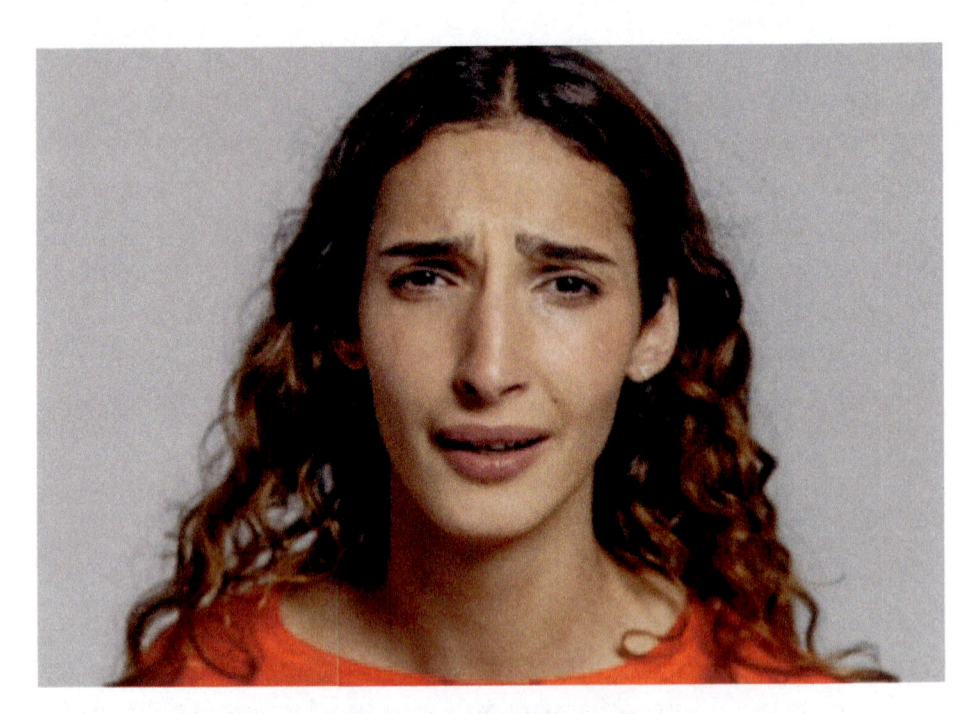

Why do I dread breastfeeding?
The causes of these feelings need sensitive treatment.

"I hate the feeling that he's on me all the time. I dread breastfeeding him. It makes me feel really low. People keep telling me that breast is best. I want to give him formula, but I feel like I am letting him down."

This was whispered to me by a tearful mum with a five week old baby at our feeding drop in. We had talked about her problems – her sore nipples, her isolation, the baby's total dependence on her. Finally, she admitted that she hated breastfeeding. Sometimes she felt an almost overwhelming low mood as she fed her baby. This seems to be one of mothers' best kept secrets, why don't we talk about it?

What could be going on? Often there are many complex and subtle issues at play. Women can dislike breastfeeding for many reasons: **physical**, **physiological**, and **psychological**.

For some women, breastfeeding is not going well. Maybe her breasts or nipples are uncomfortable, or her baby fusses, bites or tugs at the breast. There are many reasons why a mother might come to dread each feed. Most of these can be resolved or managed with some dedicated, sensitive support from a lactation professional, so if this applies to you please reach out for some help.

A little understood phenomenon is **Dysphoric Milk Ejection Reflex (D-MER)**. Mothers suffering from this physiological condition feel intense sadness and low mood from just before a breastfeed and for a couple of minutes of it starting, some experiencing it again with subsequent let downs through the feeding session. This is a chemical response in the brain to inappropriately dropping dopamine levels. The low dopamine triggers temporary but overwhelming sadness, anxiety, irritability and/or restlessness. It is not associated with physical pain or nipple trauma, or depression, and a mother cannot "think herself out of it." There is more information and support for mothers with D-MER here in the Sources of support section.

It may be that a mother is finding the **overwhelming need for her attention** just too much. There is something so intensely draining, physically and emotionally, about being entirely responsible for the wellbeing of a baby. Her baby is utterly dependent on her, demanding attention at unpredictable times of day and night with no let up. She is constantly on edge, waiting for the next time her baby needs her. Add to that sleep deprivation (which is used as a form of torture in some countries), and a woman can feel like she is losing her mind. If this is you, please recognise that this is often a totally normal response to an intense physical and emotional transition. Be kind to yourself. This is doubly true if you have had a difficult birth or another situation in your life that is compounding the issue.

Becoming a new mother can be a difficult, as well as incredible, part of your life's journey. As a society we are not good at acknowledging this and giving appropriate loving support. Please call on your friends and family, take as much help as you can, and reach out to professionals and supportive communities who can assist you. Please note that this is general advice. If a mother or her partner is worried about her overwhelming low mood, hallucinations, or thoughts of harming herself or her baby, she needs some urgent medical attention.

Some women find that **breastfeeding exacerbates their existing low mood, anxiety or depression**. The isolation, especially if worried about going out or feeding in public, makes it worse. Others find that breastfeeding can help in increasing oxytocin and bonding and lessen the symptoms.

Breastfeeding researcher Dr Alison Strube writes, *"We have much to understand about the biology of postpartum depression…when a mother is experiencing depression and anxiety symptoms, my job as a clinician is to ask what parts of her day bring her joy and peace, and what makes her symptoms worse."* This is key. For some women, breastfeeding makes her dislike her baby eight to ten times a day. We need to acknowledge this. Sometimes there are strategies that can make the experience better, and other times a mum and baby can be better served by a move to formula milk. If this resonates with you, please seek help with a lactation consultant or counsellor with of the many baby feeding organisations who can listen to you without judgement and help you to find a way forward.

I have worked with women who have experienced sexual assault or abuse, for whom birth and breastfeeding can be especially challenging. For some, the process has been handled sensitively and has been healing. In other cases, we must recognise that mum's mental health comes first, and that bottle feeding might be much more appropriate. Babies thrive on love and formula milk.

When can I have sex? Will my boobs leak?

The general advice is to wait until the postnatal blood loss is over, and then see how you feel. Breastfeeding is known to reduce libido and is sometimes associated with vaginal dryness. Some women's breasts leak milk during sex (thank you oxytocin!)

Many women ask me when it is safe to have sex after birth. Here we are talking about penis in vagina penetrative sex, acknowledging that there are many other forms of sex that may be totally appropriate whenever you want them. Blood loss has generally gone by about six weeks after the birth, so this is often seen as a time when a woman might want to consider sex again.

In the UK, women are invited for a check up with their doctor at six weeks postpartum. Some women wait for permission from their doctor to restart their sex lives. Others might try it just before the check up, so that they can address any issues at the appointment.

I know that not every woman can, or wants to, avoid sex before six weeks. However it is important to be fully healed before penetrative sex, it is not good for you to risk further damage to internal or external tissue that is healing.

For many women, six weeks is still far too early to contemplate penetrative sex. They might still be recovering from the birth, especially if it was instrumental or traumatic in some way. Physical and emotional wounds can take much longer to heal. I highly recommend seeing a specialist women's health physiotherapist for help with any perineal, pelvic floor, prolapse or other physical birth trauma.

Both mothers and partners experience profound hormonal changes after the birth of their babies, which tend to reduce sex drive. Breastfeeding mothers will find that the hormones involved can reduce libido and increase vaginal dryness. Male partners also experience a drop in testosterone after the birth of a child. Perhaps a sense of humour, glass of wine and generous application of lube can help. It is important to mention that oxytocin, released during sexual activity, also triggers milk let down in the woman. It might be necessary to have sex in a bra, with breast pads, unless dripping or spraying breastmilk is your thing!

In the UK the doctor will often talk about contraception at the six week check, as you can still get pregnant. Exclusive breastfeeding day and night can work as contraception if your baby is under six months old. However, this does mean breastfeeding on demand very frequently and not, for example, routinely giving some bottles each day or encouraging your baby to sleep overnight. There is a more detailed section on this coming up.

There is nothing physically stopping me from having sex, but I still don't want to. Is this normal?

Yes.

There are many emotional and psychological challenges that can get in the way of sex after a baby. They are not often talked about, so we will address them here.

New status as parents.

I am a huge fan of philosopher Alain de Botton, and his brilliant books on sex and love. In "How to think more about sex" Alain tells us how being a parent is not meant to be sexy. We are deeply conditioned not to find our own family members, particularly our parents, attractive

When our sexual partners are suddenly "mum" and "dad" to our new child, everything changes. This is one of the reasons why sex therapists recommend booking time away, ideally not in your own home, to reconnect and have sex. It can take physical separation from the "mummy" and "daddy" roles and space to get your heads into a sexy place instead.

Being "touched out".

It is enormously common, especially for breastfeeding mothers, to feel like they constantly have someone on them. Before the birth it is hard to imagine having someone physically and emotionally draining you all the time. As new mothers, we cannot say no to our baby, but we can say no to our partner's attempts to get affection. This is unsurprising, but hard for everyone involved. As ever, open communication and time are big healers.

Trauma.

For both partners, birth and breastfeeding can put them off sex. This is particularly true if either have been stressful, painful or a medical emergency. While it is normal, it helps to address it sensitively as soon as possible before it becomes a bigger issue than it needs to be.

It is sad to say that some women who have experienced traumatic birth, or sometimes even straightforward birth, can feel "violated." It is possible that a huge range of uncomfortable emotions could be triggered around sexual activity. This could be particularly difficult for those who have experienced any kind of abuse. It can be so hard, but please try to find a way to talk about this with your partner, trusted friends and/or a sensitive therapist, to try to find healthy ways to interact that feel comfortable for you as you heal.

Trauma is not confined to women. There are many birth partners who find it extremely difficult to get over what they have witnessed. Sometimes watching a traumatic event, and being unable to do anything to help, can be utterly overwhelming. Partners who may have previously lost someone dear to them in hospital, have experienced horror through the military or their job, or those who have witnessed trauma in any form, may find themselves taken back to these events.

I have worked with fathers who have been triggered to depression, obsessive compulsive disorders and PTSD by their experiences at the time of the birth. It can take a lot of time and understanding to help a partner to come to terms with a traumatic labour, birth and postnatal period. When all the attention seems to be on the mother, partners can feel lost and abandoned. Their needs are vitally important too.

It is not uncommon for a partner to be utterly disinterested in sex when the woman is quite up for it.

What to do about it.

Either or both partners might find that any intimacy could be construed as an invitation to have sex, and therefore consciously or subconsciously avoid it. It is said that you are twice as likely to kiss your baby as kiss each other.

The relationship books suggest you find time to be intimate and caring in ways that do not feel like they lead to sex. This might just be a head or shoulder rub, holding hands, smiling, kissing on the way past each other, snuggling on the sofa, whatever feels good to you. I cannot stress enough that talking about it is a good idea! Although it is totally understandable that in your hormonal, sleep deprived state you might not like each other at all. Reach out to friends and family who you trust and have had similar experiences. Try to hold onto the idea that this is normal and not forever.

Does breastfeeding work as contraception?

People use exclusive breastfeeding as a form of contraception, but it is not advisable if you are keen not to get pregnant. If you are exclusively breastfeeding at least every four hours during the day and up to six hours at night, with no supplemental feeds, it is unlikely (but not impossible) that you will ovulate in the first six months.

You may have heard that breastfeeding stops women from getting pregnant. In the UK, most women do not rely on breastfeeding as a form of contraception. For example, if a partner consistently gives the baby an evening bottle feed that can be enough to signal the woman's body to ovulate again.

Many women do get pregnant almost right after giving birth, even those who needed IVF last time. This might be great for you, but many healthcare professionals would advise waiting twelve months, especially after a caesarean before conceiving again.

Non-breastfeeding and combination feeding mothers will ovulate one to three months after childbirth, generally around the 45th day.

So, unless you want another baby right away, please use contraception. Barrier methods are often recommended. A coil can be fitted about six weeks post birth. Breastfeeding mothers are often told to avoid the combined contraceptive pill, patch, coil or other contraception including oestrogen. The progesterone only mini pill is considered suitable.

How can I deal with advice?

This is a tricky one. Listen to the advice and the emotion behind it, and then try to do what works for you.

"My mother in law insists that I put my baby down so that he learns to fall asleep on his own. But I don't want to, I like feeding my baby to sleep."

Parents regularly talk about the (often conflicting) advice they receive from family, friends, online and from strangers. It can feel like their feeding and parenting decisions are constantly under scrutiny. What is going on here? We are a social species, and we invest in our little ones. The advice-givers want to help, and they want validation for their own cherished ways of doing things. New mothers often seek advice, they are fearful that they are doing something wrong, and want to conform to society's ideas and expectations. While this is all understandable, it can help for a new mum to focus on her own baby, and gain confidence in her ability to make choices that are right for her and her little one.

Seeking advice.

A message I received via WhatsApp:

"To be honest, there is so much out there that I can get information overload. M is only sleeping in two hour stints, posseting a lot and is quite gassy. Some people are saying that she is not getting enough food, others say I should wind her every two minutes when I feed, someone said she is not full because the latch isn't great, but then someone else said that she is only sleeping two hour stints because she is not getting enough awake time…"

Oh my word. The poor mummy. Her baby was putting on weight, weeing and pooing frequently, just being a normal baby. As a new mum she was utterly overwhelmed with conflicting advice from well-meaning but probably not fully informed friends, family and health professionals. She had completely lost her sense of judgment and did not trust her instincts.

Just for reference, many midwives, health visitors and GPs have not had extensive training in baby feeding or sleeping. I know, because they come to me for consultations when they have their own babies.

Uninformed advice can be detrimental. Unless someone has sat with you for many hours or even days, they will not have any clue how your baby is and how your breasts work. They can only give ideas, based on their own experience, or worse drawing on limited evidence and out of date research. They may be well meaning, but they do not know your baby like you do. You know your baby best. Listen to the advice if you want to, and then try the information on for size. Does it resonate with you? Does it sound like something that might be worth a try? If not, discard it.

It is so easy for a new parent to blow with the wind, trying one thing and then the next as you read or hear about an opposing view.

Over time I hope that you will listen to your instincts, watch how your baby responds, and find something that works for you and your little one. If you are seeking information, make sure that it comes from a credible, evidence-based source. This is all part of the learning process, and as a parent it never ends.

What about unsolicited advice?

"Put that baby down, you are making a rod for your own back, she needs to learn to sleep on her own!"

"Why aren't you spacing out the time between feeds?"

"He can't still be hungry! Just give him a bottle of formula to help him sleep longer."

It might be that you are not asking for advice, but you are being given plenty of it! How can you deal with it?

As we have already said, you are the expert on your own baby. If you keep receiving advice, you might like to try any of the following suggestions from La Leche League's book "The Womanly Art of Breastfeeding."

Do reach out to your friends and support network. Surrounding yourself with people who love you and do not judge can make all the difference.

Tips for dealing with advice:

- Ask to hear their stories, without criticism from you. Many advice givers lose interest in your story as they are busy telling theirs
- Find some way to agree, and then change the subject
- Say *"thanks, we are working on it,"* which might be true
- Ask why they are telling you. *"Are you concerned? Does she look happy and healthy to you?"*
- Defuse it in advance. *"I know that David and I are feeding Liam differently from the way that you fed David. It means everything that you are so understanding."*
- Agree to differ. *"I realise that this doesn't fit with your ideas, but it is right for us."*
- Be honest. *"We find this works for us."*

Why do I feel like a failure?

This is a common feeling, but I promise that you are not a failure.

"I called an independent midwife because I am so anxious and not coping. I am ashamed. We are such failures."

This was the whispered confession of a mum of a two week old. I think she was surprised to hear me say *"Getting help is such a positive first step. Of course you are not coping, you are a brand new mum suffering from the after effects of a traumatic birth and huge blood loss. Your poor body is trying to recover, to make more blood and work on that damaged perineum. You family live in another country, so it is great that you called a professional for help."*

Where does this feeling of failure come from? When I asked her more about it, she said that this was the first time that she had not succeeded in something. All through her professional life, she followed the rules and achieved the outcome she expected. A new baby had thrown her completely. Her baby daughter would not settle in the cot, she would only sleep on her or her partner, who felt like they could not close their eyes. Everything was overwhelming.

This is a key lesson for all of us. It can be hard to admit that you are not coping, and to reach out. As new parents we need to learn how to be nurtured again, to be out of control and vulnerable and to seek and accept help from others. **This is not a failure! It is a new way of being.**

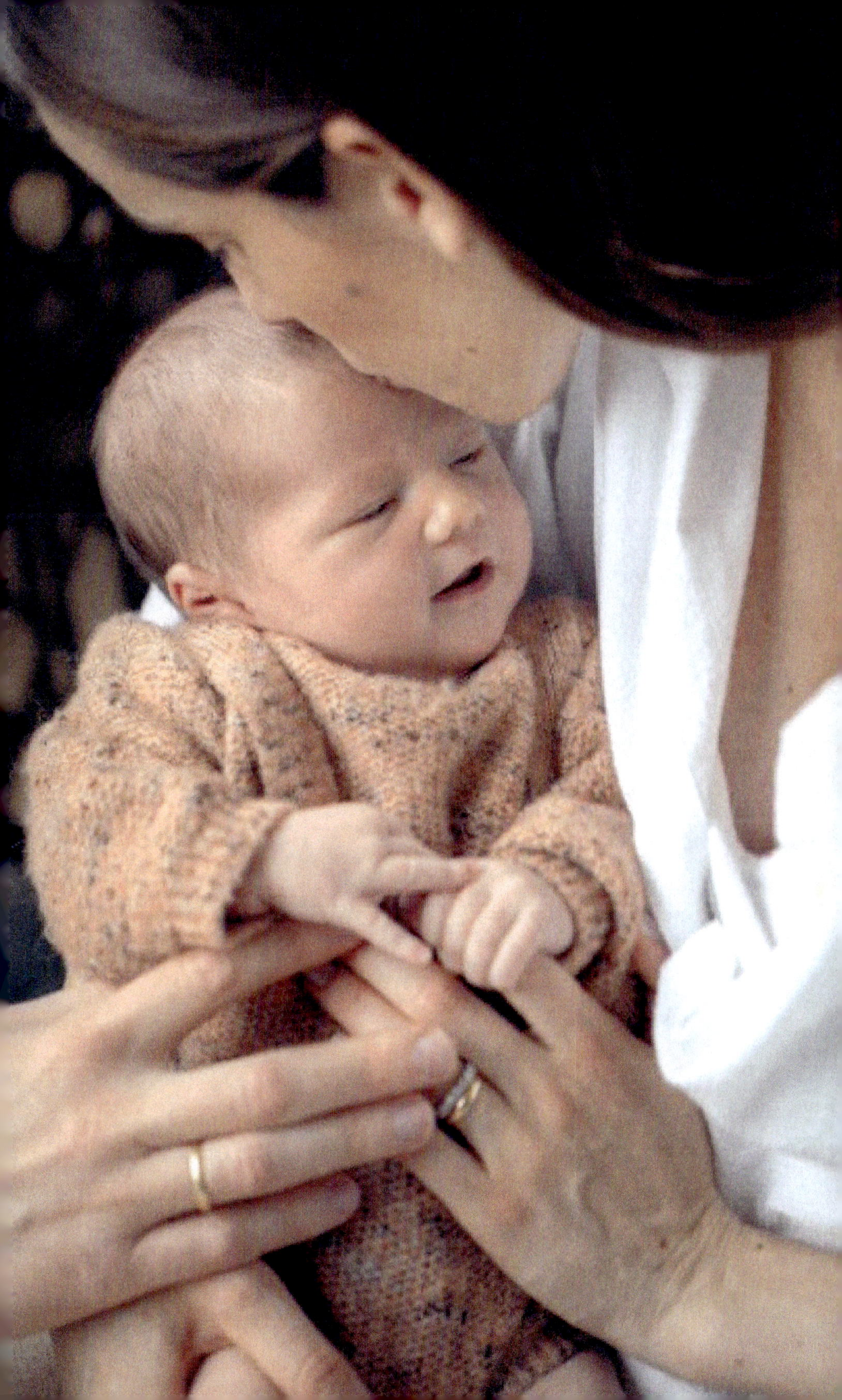

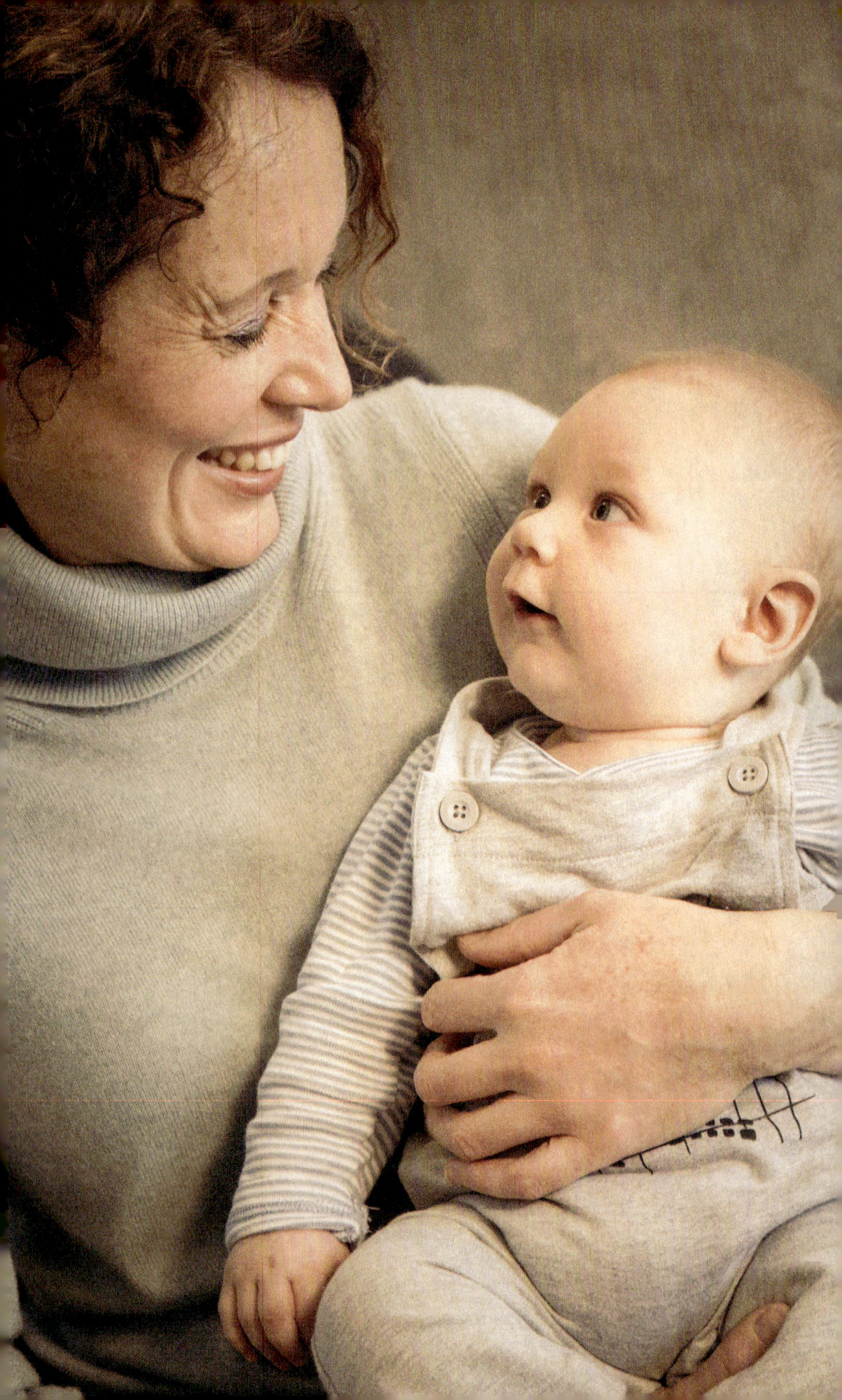

About the author

Cathy Sage is a qualified Lactation Consultant (IBCLC), NCT antenatal teacher and Breastfeeding Counsellor with over 18 years of experience. She has a busy private practice supporting new parents in London and online. Cathy volunteered for decades providing feeding support on the postnatal wards of two NHS hospitals and at her local GP surgery, and she still runs NCT breastfeeding drop-in sessions.

Cathy has prepared and supported thousands of new families in their transition to parenthood and worked with them to overcome feeding challenges. As the mother of two breast and bottle fed children, Cathy understands many of the highs and lows of the early days.

In her free time, Cathy is a keen hill walker and likes nothing better than a trip to the Lake District. She practices yoga (still can't do a headstand), reads extensively about babies, neuroscience and philosophy, and enjoys book (OK, wine) club with her friends.

Cathy firmly believes that new parents should be provided with practical, impartial evidence-based information and empathetic support. Please follow her on Instagram @cathysagebabyfeeding or visit www.cathysage.com for her latest articles and videos.

Sources of support

Breastfeeding organisations

Best beginnings
www.bestbeginnings.org.uk

Breastfeeding Network
www.breastfeedingnetwork.org.uk

La Leche League
www.laleche.org.uk

NCT
www.nct.org.uk

D-MER

D-MER
www.d-mer.org/

Evidence based resources on breast and bottle feeding

Kelly Mom, evidenced-based information from IBCLCs
www.kellymom.com

Feeding information from NCT
www.nct.org.uk/parenting/feeding

UNCEF resources on bottle feeding
www.unicef.org.uk/babyfriendly/baby-friendly-resources/bottle-feeding-resources/guide-to-bottle-feeding

Sources of support

International Board Certified Lactation Consultants

Lactation Consultants of Great Britain
Find an IBCLC near you
www.lcgb.org/find-an-ibclc

Medications

Breastfeeding Network evidence based fact sheets
www.breastfeedingnetwork.org.uk/drugs-factsheets
Searchable information at www.e-lactation.com

Mental health

Big White Wall
www.bigwhitewall.com

Mothers for mothers
www.mothersformothers.co.uk

Association for Postnatal Illness
www.apni.org

PaNDAS
www.pandasfoundation.org.uk

Mind
www.mind.org.uk/information-support/types-of-mental-healthproblems/postnatal-depression-and-perinatal-mentalhealth/about-perinatal-mental-health-problems

Action for Postpartum Psychosis
www.app-network.org

Sources of support

Milk banks

Donate or offer donor milk UKAMB
www.ukamb.org

Hearts Milk Bank
www.humanmilkfoundation.org/hearts-milk-bank

Multiples advice

Twins Trust
www.twinstrust.org

NHS

NHS pages on breastfeeding
www.nhs.uk/start-for-life/baby/feeding-your-baby/breastfeeding

Nutrition and Infant Formula

First Steps Nutrition Trust
www.firststepsnutrition.org

Premature babies

Charity Bliss
www.bliss.org.uk

Sexual abuse

Women's Aid
www.womansaid.org.uk

Government guidance
www.gov.uk/guidance/domestic-abuse-how-to-get-help

List of support services
www.mind.org.uk/information-support/guides-to-support-and-services/abuse

Sources of support

Surgery

Breastfeeding After Reduction Surgery
www.bfar.org

Therapy

Find a BACP registered therapist
www.betterhelp.com

Tongue-tie practitioners

Association of Tongue-Tie Practitioners
Find a qualified tongue-tie practitioner near you
www.tongue-tie.org

Traumatic birth recovery

Organisation for Traumatic Birth Recovery
www.traumaticbirthrecovery.com

Helplines

NCT Support Line
0300 330 0700

National Breastfeeding Helpline
0300 100 0212

Association of Breastfeeding Mothers (ABM) Helpline
0300 330 5453

La Leche League Helpline
0845 120 2918

Index

Alcohol when breastfeeding 133
Allergy
- Cow's milk protein 211
- Signs in baby 211
- Food for parents 132, 135
- Specialist formula 323-324
- Soy allergy 135, 211, 323

Autistic mother 268-270
Baby
- Allergy 211, 132, 135, 323-324
- Biting 219-223
- Colic 213
- Constipation 199
- Fussy 209-218
- Jaundice 195-198
- Lactose intolerance 212
- Mucus 194
- Nursing strike 216-218
- Poo 93-94, 198-199
- Reflux 213-215
- Sleep 114-128
- Tension 215-216
- Tongue-tie 200-208
- Weight gain 95
- Won't latch 77-79

Index

Bed sharing when breastfeeding 122-127
Birth, impact on feeding
- How to cope with a difficult start 244-245, 387
- Impact of severe blood loss 260-261
- Impact of caesarean 264-265
- Impact of instrumental birth 262-263
- Spending time in hospital 244-250, 255-257
- Unwell mother 246-247
- Unwell baby 247-254

Biting 219-213
Bleeding, postnatal 260-261
Bottle feeding 314-375
- Alternatives to bottles 351-354
- Aversion 346-347, 364-365
- Concerns 354-361
- Confusion in breastfed baby 301
- Coughing 346
- Dribbling 346
- First feeds 340-353
- Ending a feed 345
- Formula types 319-327
- Mixed feeding impact on milk supply 360-361
- Overfeeding 352
- Paced 344-345, 353, 359-360
- Partner helping at night 127-129
- Preparation by hand 332-334
- Preparation by machine 335
- Rate of milk consumption 346-347, 352-353
- Refusal 364-375

Index

- Responsive 344-345, 353, 359-360
- Signs of stress in baby during a feed 346-347
- Starting a feed 344-345, 359
- Sterilising 330-331
- Storing breastmilk 306-307
- Storing formula 338
- Teats 316-317
- Types of bottle 316-317
- Travelling 337-338
- Upright after feeds 348-349
- Volumes in bottle 342, 350-351
- Warming 337
- Water for preparation 334, 337-338
- When to start bottle feeding 35-37
- When to stop each feed 345, 352-353
- When to stop using formula 319
- Winding 88, 347-348

Breast (also see Nipples)
- Blocked ducts 151-152
- Engorgement 153-155
- Infection, Mastitis 156-158
- Pads 46
- Preparation for breastfeeding 44
- Size 19-20
- Shape 23-24, 68
- Surgery 28-32
- Symmetry 23-24

Index

Breastfeeding
- After caesarean 264
- After instrumental birth 262-263
- Autistic mother 268-270
- Biting 219-223
- Burping 88
- Clicking during feeds 203
- Cluster feeding 101-103
- Co-feeding 24-28
- Colostrum harvesting 44-45, 286-287
- Cradle 63
- Cross cradle 64
- Decisions 12-41, 385
- Duration 90-91
- Expressing 284-312
- Feeding to sleep 114
- Four to six months 111
- Frequency 82, 101-104
- Full (how you know) 87, 92-97
- Health concerns in mother 137-140
- Hiccups 203
- How it works 54-55
- Hurting 144-149
- Inducing lactation 24-28
- Koala 65
- Laid back 63
- Lying down 66
- Pillows 67
- Positions 60-70

Index

- Post partum haemorrhage 176, 260, 261
- Public/outside home 272-276
- Preparing 42-51
- Medications (see Medications)
- Mixed feeding (see Mixed Feeding)
- Multiples (see Twins and Triplets)
- Nappy output 93-94
- Naps 114-119
- Newborn – how to start 56-59, 262-264
- NICU (see NICU)
- Nose on breast 77
- Nose to nipple 75-76
- Refusal 216-218
- Relactation 189-190
- Routine 105-107
- Sexual abuse 270-271
- Sharing with partner 127-128, 296, 298
- Skin to skin 56-58, 244-245, 250-251, 256, 264
- Sling 276
- Spacing out feeds 104
- Spit up 194-195
- Straddle hold 65
- Stretching feeding interval 104
- Stopping 40, 187-189
- Surgery 28-32
- Swallowing 85, 171
- Switch sides 85-86
- Tandem 30
- Three weeks to three months 107-109

Index

- Top ups 98-99, 183-184, 234-235
- Trans parents 26-28
- Twins and triplets 224-241
- Urgent signs 96-97
- Unwell mother 137-139, 244-249, 260-262, 266-268, 304-305
- Volume 83-84
- Vomiting baby 194-195
- Vomiting mother 137
- Waking to feed 100-101
- Warning signs 95-96
- Water for baby 200
- Winding 88

Breastmilk
- Alcohol in breastmilk 133
- Caffeine in breastmilk 134-135
- Colostrum 40, 54, 286, 287, 292-294
- Composition 14-15, 307
- Defrosting 308
- Donor milk 37-38, 233, 251
- Induced lactation 24-28
- Foremilk and hindmilk 85-86
- Frozen milk quality 307
- Medications in breastmilk (see Medications)
- Milk banks 37-38
- Relactation 189-190
- Storage 279, 306-307
- Top ups 98-99, 183-184, 234-235
- Volume change over time 83-84, 175-181
- Volume required 83-84
- Warming 309

429

Index

Breast surgery 30-32
Caffeine 134-135
Caesarean 56, 66,68, 264
Cluster feeding 101-103
Colic
- Anti colic bottles 316
- Anti colic medications 213
- Anti colic formula 311
- Symptoms 213
- Treatment 213

Colostrum 40, 54, 286, 287, 292-294
Contraception when breastfeeding 140, 411-412
Cow's milk protein allergy 211
Domperidone 181
Donor milk 37-38, 233, 251
Dressing baby 58-59, 125-126
Drinking: mother
- Alcohol 133
- Caffeine 134-135
- Water 180-181

Dummies 119-121
Expressing 284-312
- Choice of expressing products 287-290
- Colostrum feeding 286-287, 292-294
- Confusion in baby 301
- Duration of session 295
- Exclusive expressing 311-312
- Frequency per day 299
- Hand expressing 291-292

Index

- Harvesting colostrum 286
- Impact on supply 298
- Maximising volume 298-299
- NICU 309-311
- One side or both 297
- Performance anxiety 298-299
- Pump and dump 302-305
- Pump types 287-290
- Reasons to express 284-285
- Rental grade pumps 290
- Returning to work 277-281
- Silicone collection systems 288
- Sterilising 305-306
- Storing expressed milk 306-307
- Times of day 296
- Volume: required 294-299
- Volume: nothing coming out 300
- When to start 291
- Wireless pumps 289-290

Feeding cues 82

Fenugreek 181

Food (mother)
- Alcohol 133
- Allergies 135-136
- Caffeine 134
- Diet 132, 135
- Dieting 136-137
- Galactagogues to boost supply 181
- Restrictions 135
- Vitamin supplements 134-135

431

Index

Forceful let down 185-186
Formula milk
- Allergy: signs 323-324
- Anti-reflux 322
- Bottled water 338
- Choice 319-327
- Comfort formula 320-321
- Composition 15, 319-327
- Constipation 199-200
- Cow's milk 319-327
- Follow on 324
- Goat's milk 323
- Halal 327
- Homemade 325
- Hungry baby 321
- Kosher 327
- Lactose free 324, 325
- Number one 329
- Organic 320
- Preparation by hand 332-334
- Preparation machines 335
- Ready to feed formula 319, 337
- Top ups 98-99, 183-184, 234-235
- Travelling 337-338
- Soy 323
- Sterilising 330-331
- Storage 338
- Vegan 327
- Vegetarian 326

Index

- Volumes 82-84, 239, 342, 350-351
- Warming 337
- Water (additional) 325

Fussy baby 209-218
- Colic 213
- Constipation 199
- Cow's Milk Protein Allergy 211
- Hunger 82, 92, 210
- Lactose intolerance 212
- Nursing strike 216-218
- Reflux 213-215
- Torticollis 215-216

Gassy baby see fussy baby

Galactagogues 181

Herpes in mother 304-305

Hospital
- Blood loss, severe 260-261
- Debrief 247, 387
- Dressing baby 58-59
- Jaundice of newborn 195-198
- Impact of caesarean 264
- Impact of instrumental birth 262-263
- Offering formula 258-259
- Postnatal bleeding – emergency signs 260-261
- Post traumatic stress disorder (PTSD) 387
- Psychiatric mother and baby unit 400-401
- Spending time 255-256

Hunger signs 82

Index

Jaundice in newborn 195-198
- Causes 196
- Diagnosis 196-197
- Treatment 197-198

Lactose
- Lactose intolerance in breastfed baby 212, 325
- Lactose free formula 324-325

Latch on breast 70-79
- Baby can't latch 77-79
- Baby's comfort 69-70
- Nose on breast 76
- Shaping the nipple 74
- Signs of good latch 72
- Signs of shallow 73
- Wide open mouth 70-71

Latch on bottle 318
- Signs of good latch 318
- Signs of baby stress when bottle feeding 346, 364-365 3676-367

Lecithin supplements 165

Legal protection 277

Let down
- DMER 405
- Flow rate 85-86, 90-91
- Forceful 185-187

Lip tie 208

Mastitis 160-161

Medications 138-140, 305
- Antibiotics 139
- Contraception 140

Index

- Contraindication to breastfeeding 137, 138, 139, 140, 266, 267, 268
- General anaesthesia 267
- Illegal drugs 266, 303
- Pain relief 138-139
- PND 398-399
- Psychosis 400-402
- Online medication checker 140

Mental health
- Advice: receiving contradictory, unhelpful or unwelcome advice 383, 412-415
- Anxiety about baby feeding 380, 383
- Baby blues 397
- Body image 381
- Dreading breastfeeding 404-405
- Exhaustion 378-382
- Feelings of failure 379-383, 416
- Guilt about feeding decisions 385
- Intrusive thoughts 402-403
- Not feeling like yourself 379-382
- Not loving your baby 386, 388-389. 408-412
- Not loving your partner 379-382,
- Perfectionism 379
- Postnatal anxiety 390-397, 402-403
- Postnatal depression: signs and treatments 398-399
- Postnatal psychosis: signs and treatments 400-402
- PTSD: signs and treatments 387
- Obsessive Compulsive Disorder 403
- Reality of life with a newborn 378-382
- Sex 407-411

Index

- Stress 390-396
- Traumatic birth: dealing with feelings 387
- Trusting instincts 383
- Unsupportive family 388-389
- Work and money 382

Milk (see breastmilk or formula milk)

Milk supply
- Boosting supply 179-181
- Causes of low supply 175-178
- Coping with breastfeeding, expressing and topping up 183-184
- Oversupply, signs, how to reduce 185-187
- Relactation 189-190
- Signs to look for 170-173
- Stopping top ups 184
- Stopping breastfeeding 187-189

Mixed feeding 98-99, 285, 298, 301
Milk tongue 161
Mucus in baby 194
Multiples 224-241
Nappy output
- Allergy/intolerance 211-212
- Constipation 199
- Early days 92-94, 172-173
- Green poo 198-199
- Oversupply 186
- Warning signs 96-97
- Urgent signs 95-96

Index

NICU
- Feeding in NICU 250-254
- Twins 224-241

Nipple
- Bleb 164-166
- Bleeding 144-149
- Confusion in baby 301
- Cracked 144-149
- Creams 149
- Flat 74-75 150-151
- Healing 149
- Inverted 20, 74, 150-151
- Piercings 20-21
- Latch 62-68, 72-74
- Shape 20, 74, 147, 150-151
- Shaping nipple for baby 74
- Shields 150-151
- Silver cups 149
- Size 74
- Sore 144-149
- Thrush 159-163
- Vasospasm 163-164

Nose (baby's nose touching breast) 77

Nursing strike 216-218

Osteopathy 215-216

Oversupply 185-186

Pumping (see expressing)
- When to pump and dump 302-305

Index

Poo
- Allergy/intolerance 211-212
- Constipation 199
- Expected in early days 92-94, 172-173
- Green poo 198-199

Post partum haemorrhage (PPH) 260-262

Postnatal depression 338-399

Postnatal psychosis 400-402

Post Traumatic Stress Disorder (PTSD) 387

Raynaud's Syndrome 147, 163-164

Reflux 213-215

Returning to work outside the home 277-281
- Childcare 281
- Expressing and storing milk 277-279, 282-301
- Feeding at night 279

Routine 105-107

Settling your baby 112-129
- Bed sharing/co-sleeping 123-127
- Cots 115-118, 122-123
- Dummies 119-121
- Feeding to sleep 114
- How much sleep required 115-116
- Putting baby down (or not) 115-118
- Nap times 117-118
- Sleep training 118-119

Sex
- Breastfeeding: leaking boobs 407
- Contraception 140, 411-412
- When to restart 407-411
- When not desired 408-411

Index

Sleep 114-128
Sleep training 118-119
Sling
- Feeding in a sling 276
- Settling 117, 393

Smoking cigarettes 303
Sore nipples 144-151
- Causes 144-148
- Treatments 149

Spit up 194-195
Sterilising 305-306, 330-331
Stopping breastfeeding 40, 48-49, 187-189
Storing milk
- Storing breastmilk 306-307
- Storing formula milk 338

Stress
- Baby stressed, how to cope 393-396
- Impact 391-392
- Maternal 390-392

Supplemental nursing system 251-253
Surgery (breast) 30, 31, 32
Tattoos 21-22
Tense neck or jaw in baby 215-216
Tongue-tie 200-208
Top ups 98-99, 183-184, 234-235
Torticollis 215-216
Trans parents 26, 27
Thrush 147, 15-163
Twins and triplets 224-241

Index

- Best start 226-235
- Coping 240
- Cup feeding 253-254
- Feeding in NICU 309-311
- Feeding positions 236-237
- Feeding together 235-237
- Finger feeding 233-234, 251-253
- Mixed feeding 238-239
- Routines 104-107
- Supplemental nursing systems 252-253
- Transitioning to breastfeeding 233-235
- Topping up 234-235
- Unique challenges in feeding multiples 226-230
- Volumes 83-84, 239, 342, 350-351

Unwell mother 137-140, 246-247, 266-268
Unwell baby 247-253
Vasospasm 163-164
Vitamin supplements baby 47, 48
Vitamin supplements mother 132
Volumes per feed 83-84, 239, 342, 350-351
Warming 309, 337
Weaning off breastmilk 40, 187-189
Wee 92-94
Weight gain 95
Wind 88-89, 186, 212-213, 347
Winding down breastfeeding 187-189

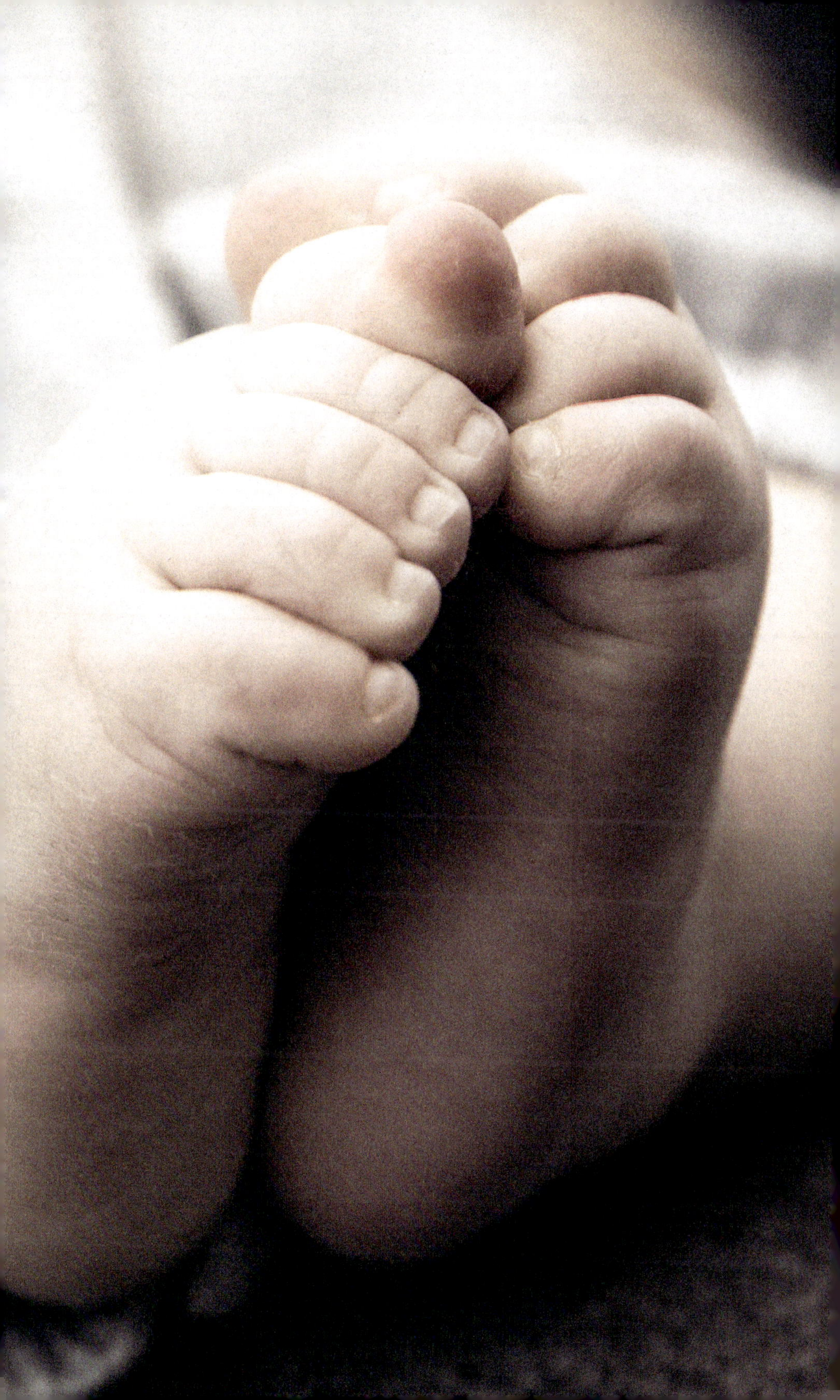

"This book is a true companion. Informed, practical guidance when you need it most, shared with warmth and empathy like a trusted friend."
Rachel Sapsford

"A godsend at 4am when everything seems to be going wrong! Every new parent should have this book on their bedside table."
Mike Daniels

"Not only is Cathy unbelievably knowledgeable, she is also personable and incredibly kind."
Kate Tutka

"Her baby led approach and calming nature is so reassuring for first time mums."
Sorcha O'Neill

"Like a favourite teacher, this book gently guides, reassures and gives you the confidence to take control of your feeding journey."
Victoria Long

"Cathy is brilliant: very knowledgeable, gentle and generous with her expertise."
Siobhan Reid

"Cathy is absolutely amazing...non-judgmental and understanding."
Matt Harris

"Cathy has a very calm, competent and empathetic approach... supporting babies and their families."
Maria Gattinoni

"I highly recommend Cathy and her book to all the expectant and new mothers in my classes at Yoga Mama. She helped my daughter to successfully breast and bottle feed her newborn."
Cherie Lathey

"Cathy has been an incredible support to many of our club members with their feeding journey. I can't recommend her highly enough."
Alex Kohansky, Bump & Baby Club

Printed in Great Britain
by Amazon